BARRIE PRICE

BUGATTI

TYPE 46 & 50

THE BIG BUGATTIS

Speedpro Series
4-Cylinder Engine Short Block High-Performance Manual – New Updated & Revised Edition (Hammill)
1275cc A-series High-performance Manual (Hammill)
Camshafts – How to Choose & Time Them For Maximum Power (Hammill)
Competition Car Datalogging Manual, The (Templeman)
Cylinder Heads, How to Build, Modify & Power Tune – Updated & Revised Edition (Burgess & Gollan)
Distributor-type Ignition Systems, How to Build & Power Tune – New 3rd Edition (Hammill)
Fast Road Car, How to Plan and Build – Revised & Updated Colour New Edition (Stapleton)
Ford SOHC 'Pinto' & Sierra Cosworth DOHC Engines, How to Power Tune – Updated & Enlarged Edition (Hammill)
Ford V8, How to Power Tune Small Block Engines (Hammill)
Harley-Davidson Evolution Engines, How to Build & Power Tune (Hammill)
Holley Carburetors, How to Build & Power Tune – Revised & Updated Edition (Hammill)
Honda Civic Type R High-Performance Manual, The (Cowland & Clifford)
Jaguar XK Engines, How to Power Tune – Revised & Updated Colour Edition (Hammill)
Land Rover Discovery, Defender & Range Rover – How to Modify Coil Sprung Models for High Performance & Off-Road Action (Hosier)
MG Midget & Austin-Healey Sprite, How to Power Tune – New 3rd Edition (Stapleton)
MGB 4-cylinder Engine, How to Power Tune (Burgess)
MGB V8 Power, How to Give Your – Third Colour Edition (Williams)
MGB, MGC & MGB V8, How to Improve – New 2nd Edition (Williams)
Mini Engines, How to Power Tune On a Small Budget – Colour Edition (Hammill)
Motorcycle-engined Racing Car, How to Build (Pashley)
Motorsport, Getting Started in (Collins)
Nissan GT-R High-performance Manual, The (Gorodji)
Nitrous Oxide High-performance Manual, The (Langfield)
Race & Trackday Driving Techniques (Hornsey)
Retro or classic car for high performance, How to modify your (Stapleton)
Rover V8 Engines, How to Power Tune (Hammill)
Secrets of Speed – Today's techniques for 4-stroke engine blueprinting & tuning (Swager)
Sportscar & Kitcar Suspension & Brakes, How to Build & Modify – Revised 3rd Edition (Hammill)
SU Carburettor High-performance Manual (Hammill)
Successful Low-Cost Rally Car, How to Build a (Young)
Suzuki 4x4, How to Modify For Serious Off-road Action (Richardson)
Tiger Avon Sportscar, How to Build Your Own – Updated & Revised 2nd Edition (Dudley)
TR2, 3 & TR4, How to Improve (Williams)
TR5, 250 & TR6, How to Improve (Williams)
TR7 & TR8, How to Improve (Williams)
V8 Engine, How to Build a Short Block For High Performance (Hammill)
Volkswagen Beetle Suspension, Brakes & Chassis, How to Modify For High Performance (Hale)
Volkswagen Bus Suspension, Brakes & Chassis for High Performance, How to Modify – Updated & Enlarged New Edition (Hale)
Weber DCOE, & Dellorto DHLA Carburetors, How to Build & Power Tune – 3rd Edition (Hammill)

Essential Buyer's Guide Series
Alfa Romeo Giulia GT Coupé (Booker)
Alfa Romeo Giulia Spider (Booker)
Audi TT (Davies)
Austin Seven (Barker)
Big Healeys (Trummel)
BMW E21 3 Series (1975-1983) (Reverente)
BMW E30 3 Series 1981 to 1994 (Hosier)
BMW GS (Henshaw)
BMW X5 (Saunders)
BSA 350 & 500 Unit Construction Singles (Henshaw)
BSA 500 & 650 Twins (Henshaw)
BSA Bantam (Henshaw)
Citroën 2CV (Paxton)
Citroën ID & DS (Heilig)
Cobra Replicas (Ayre)
Corvette C2 Sting Ray 1963-1967 (Falconer)
Fiat 500 & 600 (Bobbitt)
Ford Capri (Paxton)
Ford Escort Mk1 & Mk2 (Williamson)
Ford Mustang – First Generation 1964 to 1973 (Cook)
Ford Mustang – Fifth generation/S197 (Cook)
Ford RS Cosworth Sierra & Escort (Williamson)
Harley-Davidson Big Twins (Henshaw)
Hinckley Triumph triples & fours 750, 900, 955, 1000, 1050, 1200 – 1991-2009 (Henshaw)
Honda CBR FireBlade (Henshaw)
Honda CBR600 Hurricane (Henshaw)
Honda SOHC Fours 1969-1984 (Henshaw)
Jaguar E-Type 3.8 & 4.2-litre (Crespin)
Jaguar E-type V12 5.3-litre (Crespin)
Jaguar Mark 1 & 2 (All models including Daimler 2.5-litre V8) 1955 to 1969 (Thorley)
Jaguar S-Type – 1999 to 2007 (Thorley)
Jaguar X-Type – 2001 to 2009 (Thorley)
Jaguar XJ-S (Crespin)
Jaguar XJ6, XJ8 & XJR (Thorley)
Jaguar XK 120, 140 & 150 (Thorley)
Jaguar XK8 & XKR (1996-2005) (Thorley)
Jaguar/Daimler XJ 1994-2003 (Crespin)
Jaguar/Daimler XJ40 (Crespin)
Jaguar/Daimler XJ6, XJ12 & Sovereign (Crespin)
Lotus Seven replicas & Caterham 7: 1973-2013 (Hawkins)
Mazda MX-5 Miata (Mk1 1989-97 & Mk2 98-2001) (Crook)
Mercedes Benz Pagoda 230SL, 250SL & 280SL roadsters & coupés (Bass)
Mercedes-Benz 280-560SL & SLC (Bass)
MG Midget & A-H Sprite (Horler)
MG TD, TF & TF1500 (Jones)
MGA 1955-1962 (Crosier)
MGB & MGB GT (Williams)
MGF & MG TF (Hawkins)
Mini (Paxton)
Morris Minor & 1000 (Newell)
Moto Guzzi 2-valve big twins (Falloon)
New Mini (Collins)
Norton Commando (Henshaw)
Peugeot 205 GTI (Blackburn)
Porsche 911 (964) (Streather)
Porsche 911 (993) (Streather)
Porsche 911 (996) (Streather)
Porsche 911 Carrera 3.2 (Streather)
Porsche 911SC (Streather)
Porsche 924 – All models 1976 to 1988 (Hodgkins)
Porsche 928 (Hemmings)
Porsche 930 Turbo & 911 (930) Turbo (Streather)
Porsche 944 (Higgins)
Porsche 986 Boxster (Streather)
Porsche 987 Boxster & Cayman (Streather)
Rolls-Royce Silver Shadow & Bentley T-Series (Bobbitt)
Subaru Impreza (Hobbs)
Triumph Stag (Mort)
Triumph TR6 (Williams)
Triumph TR7 & TR8 (Williams)
Volvo 700/900 Series (Beavis)
VW Beetle (Cservenka & Copping)
VW Bus (Cservenka & Copping)
VW Golf GTI (Cservenka & Copping)

Those Were The Days ... Series
Alpine Trials & Rallies 1910-1973 (Pfundner)
American 'Independent' Automakers – AMC to Willys 1945 to 1960 (Mort)
American Station Wagons – The Golden Era 1950-1975 (Mort)
American Trucks of the 1950s (Mort)
American Trucks of the 1960s (Mort)
American Woodies 1928-1953 (Mort)
Anglo-American Cars from the 1930s to the 1970s (Mort)
Austerity Motoring (Bobbitt)
Austins, The last real (Peck)
Brighton National Speed Trials (Gardiner)
British and European Trucks of the 1970s (Peck)
British Drag Racing – The early years (Pettitt)
British Lorries of the 1950s (Bobbitt)
British Lorries of the 1960s (Bobbitt)
British Touring Car Racing (Collins)
British Police Cars (Walker)
British Woodies (Peck)
Café Racer Phenomenon, The (Walker)
Don Hayter's MGB Story – The birth of the MGB in MG's Abingdon Design & Development Office (Hayter)
Drag Bike Racing in Britain – From the mid '60s to the mid '80s (Lee)
Endurance Racing at Silverstone in the 1970s & 1980s (Parker)
Hot Rod & Stock Car Racing in Britain in the 1980s (Neil)
Last Real Austins 1946-1959, The (Peck)
MG's Abingdon Factory (Moylan)
Motor Racing at Brands Hatch in the Seventies (Parker)
Motor Racing at Brands Hatch in the Eighties (Parker)
Motor Racing at Crystal Palace (Collins)
Motor Racing at Goodwood in the Sixties (Gardiner)
Motor Racing at Nassau in the 1950s & 1960s (O'Neil)
Motor Racing at Oulton Park in the 1960s (McFadyen)
Motor Racing at Oulton Park in the 1970s (McFadyen)
Motor Racing at Thruxton in the 1970s (Grant-Braham)
Motor Racing at Thruxton in the 1980s (Grant-Braham)
Superprix – The Story of Birmingham Motor Race (Page & Collins)
Three Wheelers (Bobbitt)

Biographies
A Chequered Life – Graham Warner and the Chequered Flag (Hesletine)
Amédée Gordini ... a true racing legend (Smith)
André Lefebvre, and the cars he created at Voisin and Citroën (Beck)
Chris Carter at Large – Stories from a lifetime in motorcycle racing (Carter & Skelton)
Cliff Allison, The Official Biography of – From the Fells to Ferrari (Gauld)
Edward Turner – The Man Behind the Motorcycles (Clew)
Driven by Desire – The Desiré Wilson Story
First Principles – The Official Biography of Keith Duckworth (Burr)
Inspired to Design – F1 cars, Indycars & racing tyres: the autobiography of Nigel Bennett (Bennett)
John Chatham – 'Mr Big Healey' – The Official Biography (Burr)
The Lee Noble Story (Wilkins)
Mason's Motoring Mayhem – Tony Mason's hectic life in motorsport and television (Mason)
Raymond Mays' Magnificent Obsession (Apps)
Pat Moss Carlsson Story, The – Harnessing Horsepower (Turner)
Tony Robinson – The biography of a race mechanic (Wagstaff)
Virgil Exner – Visioneer: The Official Biography of Virgil M Exner Designer Extraordinaire (Grist)

General
1½-litre GP Racing 1961-1965 (Whitelock)
AC Two-litre Saloons & Buckland Sportscars (Archibald)
Alfa Romeo 155/156/147 Competition Touring Cars (Collins)
Alfa Romeo Giulia Coupé GT & GTA (Tipler)
Alfa Romeo Montreal – The dream car that came true (Taylor)
Alfa Romeo Montreal – The Essential Companion (Classic Reprint of 500 copies) (Taylor)
Alfa Tipo 33 (McDonough & Collins)
Alpine & Renault – The Development of the Revolutionary Turbo F1 Car 1968 to 1979 (Smith)
Alpine & Renault – The Sports Prototypes 1963 to 1969 (Smith)
Alpine & Renault – The Sports Prototypes 1973 to 1978 (Smith)
Anatomy of the Works Minis (Moylan)
Armstrong-Siddeley (Smith)
Art Deco and British Car Design (Down)
Autodrome (Collins & Ireland)
Autodrome 2 (Collins & Ireland)
Automotive A-Z, Lane's Dictionary of Automotive Terms (Lane)
Automotive Mascots (Kay & Springate)
Bahamas Speed Weeks, The (O'Neil)
Bentley Continental, Corniche and Azure (Bennett)
Bentley MkVI, Rolls-Royce Silver Wraith, Dawn & Cloud/Bentley R & S-Series (Nutland)
Bluebird CN7 (Stevens)
BMC Competitions Department Secrets (Turner, Chambers & Browning)
BMW 5-Series (Cranswick)
BMW Z-Cars (Taylor)
BMW Boxer Twins 1970-1995 Bible, The (Falloon)
BMW Cafe Racers (Cloesen)
BMW Custom Motorcycles – Choppers, Cruisers, Bobbers, Trikes & Quads (Cloesen)
BMW – The Power of M (Vivian)
Bonjour – Is this Italy? (Turner)
British 250cc Racing Motorcycles (Pereira)
British at Indianapolis, The (Wagstaff)
British Cars, The Complete Catalogue of, 1895-1975 (Culshaw & Horrobin)
British Custom Motorcycles – The Brit Chop – choppers, cruisers, bobbers & trikes (Cloesen)
BRM – A Mechanic's Tale (Salmon)
BRM V16 (Ludvigsen)
BSA Bantam Bible, The (Henshaw)
BSA Motorcycles – the final evolution (Jones)
Bugatti Type 40 (Price)
Bugatti 46/50 Updated Edition (Price & Arbey)
Bugatti T44 & T49 (Price & Arbey)
Bugatti 57 2nd Edition (Price)
Bugatti Type 57 Grand Prix – A Celebration (Tomlinson)
Caravan, Improve & Modify Your (Porter)
Caravans, The Illustrated History 1919-1959 (Jenkinson)
Caravans, The Illustrated History From 1960 (Jenkinson)
Carrera Panamericana, La (Tipler)
Chrysler 300 – America's Most Powerful Car 2nd Edition (Ackerson)
Chrysler PT Cruiser (Ackerson)
Citroën DS (Bobbitt)
Classic British Car Electrical Systems (Astley)
Cobra – The Real Thing! (Legate)
Competition Car Aerodynamics 3rd Edition (McBeath)
Concept Cars, How to illustrate and design (Dewey)
Cortina – Ford's Bestseller (Robson)
Coventry Climax Racing Engines (Hammill)
Daily Mirror 1970 World Cup Rally 40, The (Robson)
Daimler SP250 New Edition (Long)
Datsun Fairlady Roadster to 280ZX – The Z-Car Story (Long)
Dino – The V6 Ferrari (Long)
Dodge Challenger & Plymouth Barracuda (Grist)
Dodge Charger – Enduring Thunder (Ackerson)
Dodge Dynamite! (Grist)
Dorset from the Sea – The Jurassic Coast from Lyme Regis to Old Harry Rocks photographed from its best viewpoint (Belasco)
Dorset from the Sea – The Jurassic Coast from Lyme Regis to Old Harry Rocks photographed from its best viewpoint (souvenir edition) (Belasco)
Draw & Paint Cars – How to (Gardiner)
Drive on the Wild Side, A – 20 Extreme Driving Adventures From Around the World (Weaver)
East German Motor Vehicles in Pictures (Suhr/Weinreich)
Fast Ladies – Female Racing Drivers 1888 to 1970 (Bouzanquet)
Fate of the Sleeping Beauties, The (op de Weegh/Hottendorff/op de Weegh)
Ferrari 288 GTO, The Book of the (Sackey)
Ferrari 333 SP (O'Neil)
Fiat & Abarth 124 Spider & Coupé (Tipler)
Fiat & Abarth 500 & 600 – 2nd Edition (Bobbitt)
Fiats, Great Small (Ward)
Ford Cleveland 335-Series V8 engine 1970 to 1982 – The Essential Source Book (Hammill)
Ford F100/F150 Pick-up 1948-1996 (Ackerson)
Ford F150 Pick-up 1997-2005 (Ackerson)
Ford GT – Then, and Now (Streather)
Ford GT40 (Legate)
Ford Model Y (Roberts)
Ford Small Block V8 Racing Engines 1962 to 1970 – The Essential Source Book (Hammill)
Ford Thunderbird From 1954, The Book of the (Long)
Formula 5000 Motor Racing, Back then ... and back now (Lawson)
Forza Minardi! (Vigar)
France: the essential guide for car enthusiasts – 200 things for the car enthusiast to see and do (Parish)
From Crystal Palace to Red Square – A Hapless Biker's Road to Russia (Turner)
Funky Mopeds (Skelton)
Grand Prix Ferrari – The Years of Enzo Ferrari's Power, 1948-1980 (Pritchard)
Grand Prix Ford – DFV-powered Formula 1 Cars (Pritchard)
GT – The World's Best GT Cars 1953-73 (Dawson)
Hillclimbing & Sprinting – The Essential Manual (Short & Wilkinson)
Honda NSX (Long)
Inside the Rolls-Royce & Bentley Styling Department – 1971 to 2001 (Hull)
Intermeccanica – The Story of the Prancing Bull (McCredie & Reisner)
Italian Cafe Racers (Cloesen)
Italian Custom Motorcycles (Cloesen)
Jaguar, The Rise of (Price)
Jaguar XJ 220 – The Inside Story (Moreton)
Jaguar XJ-S, The Book of the (Long)
Karmann-Ghia Coupé & Convertible (Bobbitt)
Lamborghini Miura Bible, The (Sackey)
Lamborghini Urraco, The Book of the (Landsem)
Lambretta Bible, The (Davies)
Lancia 037 (Collins)
Lancia Delta HF Integrale (Blaettel & Wagner)
Land Rover Series III Reborn (Porter)
Land Rover, The Half-ton Military (Cook)
Laverda Twins & Triples Bible 1968-1986 (Falloon)
Lea-Francis Story, The (Price)
Le Mans Panoramic (Ireland)
Lexus Story, The (Long)
Little book of microcars, the (Quellin)
Little book of smart, the – New Edition (Jackson)
Little book of trikes, the (Quellin)
Lola – The Illustrated History (1957-1977) (Starkey)
Lola – All the Sports Racing & Single-seater Racing Cars 1978-1997 (Starkey)
Lola T70 – The Racing History & Individual Chassis Record – 4th Edition (Starkey)
Lotus 49 (Oliver)
Marketingmobiles, The Wonderful Wacky World of (Hale)
Maserati 250F In Focus (Pritchard)
Mazda MX-5/Miata 1.6 Enthusiast's Workshop Manual (Grainger & Shoemark)
Mazda MX-5/Miata 1.8 Enthusiast's Workshop Manual (Grainger & Shoemark)
The book of the Mazda MX-5 Miata – The 'Mk1' NA-series 1988 to 1997 (Long)
Mazda MX-5 Miata Roadster (Long)
Maximum Mini (Booij)
Meet the English (Bowie)
Mercedes-Benz SL – R230 series 2001 to 2011 (Long)
Mercedes-Benz SL – W113-series 1963-1971 (Long)
Mercedes-Benz SL & SLC – 107-series 1971-1989 (Long)
Mercedes-Benz SLK – R170 series 1996-2004 (Long)
Mercedes-Benz SLK – R171 series 2004-2011 (Long)
Mercedes-Benz W123-series – All models 1976 to 1986 (Long)
MGA (Price Williams)
MGB & MGB GT– Expert Guide (Auto-doc Series) (Williams)
MGB Electrical Systems Updated & Revised Edition (Astley)
Micro Caravans (Jenkinson)
Micro Trucks (Mort)
Microcars at Large! (Quellin)
Mini Cooper – The Real Thing! (Tipler)
Mini Minor to Asia Minor (West)
Mitsubishi Lancer Evo, The Road Car & WRC Story (Long)
Montlhéry, The Story of the Paris Autodrome (Boddy)
Morgan Maverick (Lawrence)
Morgan 3 Wheeler – back to the future!, The (Dron)
Morris Minor, 60 Years on the Road (Newell)
Moto Guzzi Sport & Le Mans Bible, The (Falloon)
Motor Movies – The Posters! (Veysey)
Motor Racing – Reflections of a Lost Era (Carter)
Motor Racing – The Pursuit of Victory 1930-1962 (Carter)
Motor Racing – The Pursuit of Victory 1963-1972 (Wyatt/Sears)
Motor Racing Heroes – The Stories of 100 Greats (Newman)
Motorcycle Apprentice (Cakebread)
Motorcycle GP Racing in the 1960s (Pereira)
Motorcycle Road & Racing Chassis Designs (Noakes)
Motorhomes, The Illustrated History (Jenkinson)
Motorsport In colour, 1950s (Wainwright)
MV Agusta Fours, The book of the classic (Falloon)
N.A.R.T. – A concise history of the North American Racing Team 1957 to 1983 (O'Neil)
Nissan 300ZX & 350Z – The Z-Car Story (Long)
Nissan GT-R Supercar: Born to race (Gorodji)
Northeast American Sports Car Races 1950-1959 (O'Neil)
Nothing Runs – Misadventures in the Classic, Collectable & Exotic Car Biz (Slutsky)
Pontiac Firebird (Cranswick)
Porsche Boxster (Long)
Porsche 356 (2nd Edition) (Long)
Porsche 908 (Födisch, Neßhöver, Roßbach, Schwarz & Roßbach)
Porsche 911 Carrera – The Last of the Evolution (Corlett)
Porsche 911R, RS & RSR, 4th Edition (Starkey)
Porsche 911, The Book of the (Long)
Porsche 911SC 'Super Carrera' – The Essential Companion (Streather)
Porsche 914 & 914-6: The Definitive History of the Road & Competition Cars (Long)
Porsche 924 (Long)
The Porsche 924 Carreras – evolution to excellence (Smith)
Porsche 928 (Long)
Porsche 944 (Long)
Porsche 964, 993 & 996 Data Plate Code Breaker (Streather)
Porsche 993 'King Of Porsche' – The Essential Companion (Streather)
Porsche 996 'Supreme Porsche' – The Essential Companion (Streather)
Porsche Racing Cars – 1953 to 1975 (Long)
Porsche Racing Cars – 1976 to 2005 (Long)
Porsche – The Rally Story (Meredith)
Porsche: Three Generations of Genius (Meredith)
Preston Tucker & Others (Linde)
RAC Rally Action! (Gardiner)
RACING COLOURS – MOTOR RACING COMPOSITIONS 1908-2009 (Newman)
Racing Line – British motorcycle racing in the golden age of the big single (Guntrip)
Rallye Sport Fords: The Inside Story (Moreton)
Roads with a View – England's greatest views and how to find them by road (Corfield)
Rolls-Royce Silver Shadow/Bentley T Series Corniche & Camargue – Revised & Enlarged Edition (Bobbitt)
Rolls-Royce Silver Spirit, Silver Spur & Bentley Mulsanne 2nd Edition (Bobbitt)
Runways & Racers (O'Neil)
Russian Motor Vehicles – Soviet Limousines 1930-2003 (Kelly)
Russian Motor Vehicles – The Czarist Period 1784 to 1917 (Kelly)
RX-7 – Mazda's Rotary Engine Sportscar (Updated & Revised New Edition) (Long)
Sleeping Beauties USA – abandoned classic cars & trucks (Marek)
SM – Citroën's Maserati-engined Supercar (Long & Claverol)
Standard Motor Company, The Book of the
Subaru Impreza: The Road Car And WRC Story (Long)
Supercar, How to Build your own (Thompson)
Tales from the Toolbox (Oliver)
Tatra – The Legacy of Hans Ledwinka, Updated & Enlarged Collector's Edition of 1500 copies (Margolius & Henry)
Toleman Story, The (Hilton)
Toyota Celica & Supra, The Book of Toyota's Sports Coupés (Long)
Toyota MR2 Coupés & Spyders (Long)
Triumph TR6 (Kimberley)
Two Summers – The Mercedes-Benz W196R Racing Car (Ackerson)
Unraced (Collins)
VW Beetle Cabriolet – The full story of the convertible Beetle (Bobbitt)
VW Beetle – The Car of the 20th Century (Copping)
VW Bus – 40 Years of Splitties, Bays & Wedges (Copping)
VW Bus Book, The (Bobbitt)
VW Golf: Five Generations of Fun (Copping & Cservenka)
VW – The Air-cooled Era (Copping)
You & Your Jaguar XK8/XKR – Buying, Enjoying, Maintaining, Modifying – New Edition (Thorley)
Which Oil? – Choosing the right oils & greases for your antique, vintage, veteran, classic or collector car (Michell)
Works Minis, The Last (Purves & Brenchley)

www.veloce.co.uk

For post publication news, updates and amendments relating to this book please visit www.veloce.co.uk/books/V4872

First published in 1995. Second edition first published in 2000 by Veloce Publishing Limited, Veloce House, Parkway Farm Business Park, Middle Farm Way, Poundbury, Dorchester DT1 3AR, England. Fax 01305 268864 / e-mail info@veloce.co.uk / web www.veloce.co.uk or www.velocebooks.com. Reprinted September 2015. [ISBN: 978-1-845848-72-9 UPC: 6-36847-04872-3].
 British Library Cataloguing in Publication Data – A catalogue record for this book is available from the British Library. Typesetting, design and page make-up all by Veloce Publishing Ltd on Apple Mac. Printed in India by Replika Press.

BARRIE PRICE

BUGATTI
TYPE 46 & 50

THE BIG BUGATTIS

VELOCE PUBLISHING
THE PUBLISHER OF FINE AUTOMOTIVE BOOKS

ACKNOWLEDGEMENTS & PUBLISHER'S NOTE

In writing a work like this, one is dependent on help from so many quarters that full acknowledgement to all concerned is impossible. I am particularly grateful, however, to the following, without whom the work could not have been done: Uwe Hucke, M. Delerme & M. Garnier of the National Motor Museum (Mulhouse), Richard Day, Tim Curr & Julie Bate of the Bugatti Trust, Ian Preston.

Photographs frequently come down over the years in a form which makes it difficult or impossible to accurately trace their origin. I therefore extend my thanks to all the photographers and artists whose work appears in these pages.

Barrie Price

Publisher's note. A number of the photographs and line illustrations used in this book are of poor original quality or have, over the years, become damaged or marked: such faults are reflected in the quality of reproduction. However, as this book represents an extraordinary feat of research and is the most complete pictorial record of the Bugatti Types 46 & 50, related models and associated coachwork ever published, it was decided to include such pictures for the sake of completeness.

Veloce

BUGATTI

46/50

CONTENTS

I

INTRODUCTION TO THE TYPE 46 & 50 BUGATTIS

Ettore Bugatti had in mind, from the early days prior to 1914, that he would one day build the ultimate large luxury car. This intention clearly indicates that Bugatti was convinced, even as a young man, that he was capable of anything in the field of engineering. Subsequent events proved this notion nearly correct although he was greatly assisted by the professional engineers and designers he employed in later years as his firm grew in size and stature.

We can categorise the motor car designs of Bugatti into four stages (discounting the early formative years in the first decade of the 20th century).

A significant engine design change heralds each of these phases. The new feature of engine design concerned was, as far as I know, always novel and unique to Bugatti - illustrating clearly that he was completely free and unfettered by convention in his design philosophy, no doubt due to his lack of a formal technical education. He thus attempted solutions for engineering problems from which professional engineers would shrink and, furthermore, he succeeded, often brilliantly!

We can categorise the stages as follows:

STAGE 1: THE "BRESCIA" ERA

The unique feature of these engines concerned the transmission of motion between the single overhead camshaft and each valve, which was by way of a curved 'banana' tappet running in a bed of white metal. The very first *"Pur Sang"* Bugatti of 1909 embodied this scheme, which continued until the last cars of this type were completed in 1925. The most significant change during this period concerned the adoption of four valves per cylinder in 1914.

These machines were brilliantly successful as high performance small cars, and firmly established the Bugatti reputation. Bear in mind that *"Le Patron"* was only 27 years old when he conceived this design and commenced production as a car manufacturer.

STAGE 2: THREE-VALVE NARROW BLOCK GENERATION

These were the most prolific and popular types which exemplified the individuality of the Bugatti car. This engine layout featured three vertical valves per cylinder and embodied a non-detachable cylinder head (as with all Bugatti types): all this was compressed into an extremely narrow engine form with all faces square and machined all over, thus providing a unique and most pleasing - one might even say spectacular - appearance. Although water space around the valves was limited, there was a gain in this respect due to the combined cylinder head and block which avoided the two decks, a gasket and a 'forest' of bosses for cylinder head studs which would have been necessary with a conventional design.

The first incarnation of this engine type was seen in the aero-engine laid out in 1916 and subsequently developed and built in America by the Duesenberg Corporation.

All car engines, commencing with the stillborn 3-litre Type 28 of 1921

Deepening the one-piece cylinder block to carry the crankshaft journals produced a very rigid engine, and represented a feat of engineering. This is the Type 46's cylinder block.

and ending with the 3.3-litre Type 49 current from 1931-34, followed this practice. The design, where concerned with 8-cylinder cars, utilised two separate cylinder blocks, each containing 4 cylinders.

STAGE 3: ONE-PIECE ENGINES

The third stage in engine layout evolved by Bugatti - and that which concerns us in this volume - carried the unitary construction principle one step further.

Bugatti now added the main bearings for the crankshaft onto the bottom of the cylinder casting resulting in a single main iron casting for a straight eight engine: a veritable *tour de force* for the Bugatti foundry and providing an engine of outstanding rigidity compared with usual standards of the time. The smoothness of running was also of a very high order, no doubt aided by the stiffness of this type of construction, which appears unique.

The crankshaft was surrounded by an aluminium 'box', together with various plates, all machined to close tolerances and thus oil-tight (if assembled to Bugatti standards of care and cleanliness).

This type of construction was initially seen in another aero-engine, given designation Type 34 and evolved in 1925.

The first car engine of one-piece configuration was the enormous Type 41 "Royale," initially of 125mm bore x 150mm stroke (14,760cc) but later reduced in stroke by 20mm to 130mm and thus 12,765cc. The first drawing that we can trace (Con Rod) is dated 15.11.28, although there would, no doubt, be some preliminary layout drawings beforehand.

The Royale has been written up so many times that we do not need to dwell upon it in this book except to say that, whilst the chassis followed standard Bugatti practice, the transmission differed in one important respect: the gearbox was placed in unit with the rear axle.

We cannot be certain why Bugatti also chose this transmission layout for the Types 46 and 50. Placing the gearbox in unit with the rear axle caused an increase in unsprung weight, complication of the gear change mechanism and the propshaft was forced to follow engine speed in all gears increasing inertia and making for a slower gear change. Bugatti may have been concerned with the problem of torque reaction through the body and chassis structure which a large engine in bottom gear will produce. Indeed, this would have been a serious problem with the enormous Royale, however, the first Bugatti design to incorporate this transmission layout was the stillborn 3-litre Type 28 of 1921: a model for which torque reaction problems would not be a serious issue. Perhaps Bugatti was worried about bottom gear loads through the propshaft, or felt that gearbox noise from the straight cut spur gears then in use would be better insulated from the passenger compartment if they were placed in the rear axle.

The most likely explanation is that the rear axle-mounted gearbox was simply a case of Bugatti individuality and original thinking although the system had been adapted occasionally by others, notably by the British firms of Singer and A.C.

STAGE 4: THE FINAL GENERATION

The 'final' generation of Bugatti engines, drawn up in 1932, was the Type 57, followed by variants under

Right: The Type 46's gearbox located in the rear axle may have created more problems than it solved.

Below: Bugatti's ingenious method of preventing self-servo action of the brakes caused by axle rotation under braking.

type numbers 59, 64 and, finally, 101.

It was a straight eight engine once more, typically Bugatti in appearance and finish but totally different in design, having little in common with previous models.

The main changes concerned the timing gears, which were situated at the rear of the engine, close to the flywheel, thus cushioned from crankshaft torsional vibrations - a good point - whilst the twin camshafts actuated the valves by way of an oscillating finger between each valve and cam lobe. A scheme inferior to the previous, direct-acting 'bucket' tappet of Types 50 and 51.

We can, perhaps, discount a further generation of engines conceived during and shortly after the Second World War. None were to enter production, were undeveloped at the time of Ettore's death and did not seem to posses any particular points of merit.

The magic has disappeared!

TYPE 46

Bugatti decided to bridge the gap between the excellent 3-litre Type 44 touring car and the gigantic Royale with a 5-litre luxury car designated Type 46.

The first production drawing (camshaft) is dated January 28 1929 and production began towards the end of that year. The car was launched at the Paris show in October, while the first delivery to a private customer appears to be chassis 46121 on November 16 1929. However, the factory records for the Type 46 are scanty and further research is needed.

The layout of the Type 46 did, in fact, follow very closely that of the Type 41, including the transmission arrangements aforementioned, incorporating a three-speed gearbox in the rear axle. Rudge wire wheels were specified in place of the magnificent aluminium type designed for the Royale, although another elegant wheel was conceived for the later Type 50 which became available as an option for the Type 46. Several earlier cars were retrospectively converted.

Considerable thought was given to the brake cable layout in order to neutralise the self-servo action of the front brakes caused by axle twist and always exaggerated with large diameter drums. A cable route passing over the kingpin centre and then turning at right angles to a further right-angle on the chassis was utilised, a scheme first seen on the Royale and again in the second series Type 57.

A further novel feature concerned a flexible mounting for the flywheel, a detail rarely - if ever - seen before or since and no doubt of assistance in breaking up torsional vibrations.

The engine dimensions were 81mm bore x 130mm stroke (5360cc)

Borrowing from Miller design philosophy, the new Bugatti twin cam design with supercharger was a superb power unit for the new T50.

and a single, five-jet Smith-Bariquand carburettor was specified. It is interesting to note that Bugatti built his own carburettor for the Royale, presumably because of the impossibility of finding a proprietary unit with chokes large enough for the 12.75-litre engine.

A small supercharger was added at a later date (models so equipped being designated Type 46S) which added to the engine's already impressive torque. There is no doubt that the Type 46 offered a standard of performance, ride and handling superior to its competitors, although the D8 Delage which followed shortly afterwards was to offer serious competition and did, in fact, sell in much greater numbers: over 2000 examples between 1930 and 1934 and the majority between 1930 and 1931. The Delage was, however, somewhat cheaper at £700 compared with the Bugatti at £1400.

TYPE 50

The next development of the one-piece engine theme was a conversion to a hemispherical combustion chamber, twin camshaft, two-valve cylinder head which was copied from the Miller racing cars that Bugatti bought to study. The new engine was designed prior to a similar conversion of the racing Type 35B which became Type 51. The first drawing, for a tappet bucket, is dated July 31 1930.

The Type 50 engine dimensions were changed, compared with the Type 46, at 86mm bore x 107mm stroke, thus squarer and smaller at 4972cc. This bore/stroke ratio would assist in reaching higher rpm and all Type 50 engines were supercharged with a large Roots-type instrument of typical Bugatti design.

200bhp was claimed at 4000rpm; indeed, it could hardly be less, bearing in mind the efficient design, good thermodynamics and fairly high compression. Certainly no other production luxury car in 1930 had such an advanced 'racing car' specification, except the Duesenberg SJ from America.

The only other design change concerned the elegant aluminium wheels mentioned already, together with a shorter chassis at 3.1 metres wheelbase, compared with the 3.5 metres of the Type 46. Later, the longer

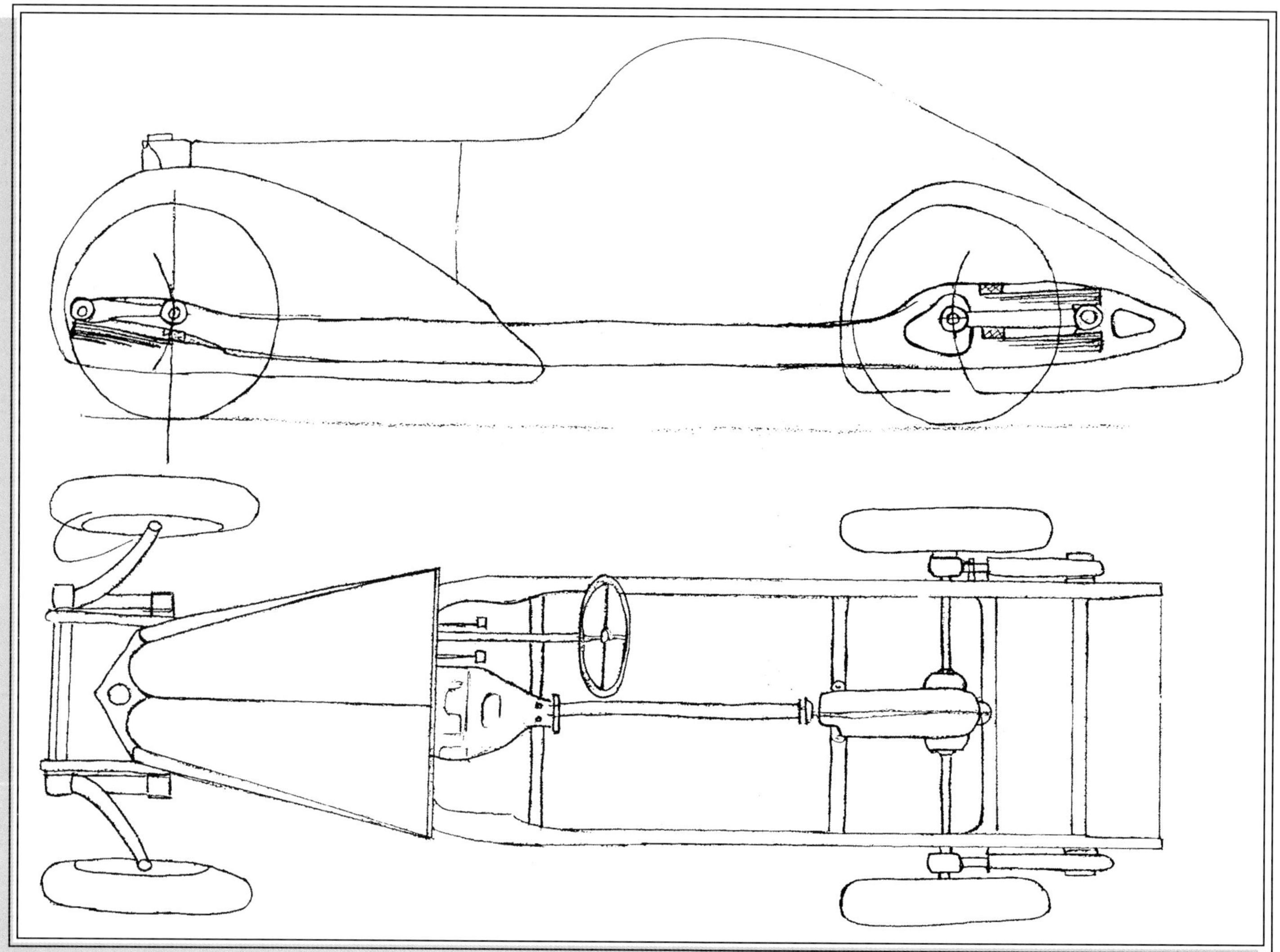

Plan and side elevation drawings, dated October 1934, of Type 50 Sports. Only one chassis was built. Note the independent suspension.

chassis was offered as an option.

The first car delivered - chassis no. 50116, engine no. 1 - went to Paris agent Lamberjack on October 11 1930. Chassis no. 50118 is recorded as being sold to Fayol of Vichy on March 13 1930, but this must surely be a mistake for we see just one other delivery in 1930: chassis no. 50121 to Bucar of Zurich on December 1. Chassis no. 50122, also to Bucar, is dated December 10 1931, which is almost certainly another misprint that I believe should read December 10 1930!

The final delivery, of chassis no. 50176 to Guerin Grenoville of Asnières, Paris, occurred on December 5 1934

1929 Paris Salon show chassis on display: note the exhaust stubs.

Ian Preston in chassis no. 50139 on the startline at a Prescott hillclimb.

and the final chassis no. is 50177, a Le Mans team car owned at the time of writing by Miles Coverdale.

TYPE 50S

A further development concerned the design of a new super sports car known as the Type 50S. One such chassis was built at Molsheim but was not completed as the project was abandoned. It was no doubt altogether too grand and expensive for the difficult market that existed in 1934. Furthermore, the entire effort of the car design office was taken up with the Type 57 family. It is greatly to the credit of Uwe Hucke that he has been able to complete this interesting car.

An important change incorporated in the Type 50S concerns the adoption of an unusual type of hydraulic brake, in which the wheel cylinders are situated on the outside of the backplates and operate a standard mechanical cam and lever via a pushrod.

I have also seen a drawing of a V-fronted radiator cowl intended for this car, together with an outline sketch of a closed body clearly anticipating the Atlantic style adopted for the 57S.

TYPES 53 & 54

The Type 50 engine was utilised in tuned-up form for the Type 53 four-wheel-drive hillclimb car (two examples) and the conventional and extremely fast Grand Prix Type 54 (approximately five examples), 300bhp and more in a rigid axle chassis and cable brakes, was really tempting providence; one car was timed on the long straight at the Avus track in Berlin at 160mph! Certainly two drivers were killed - Count Czaykowski and Prince Lobkowitz - but other factors were involved. At speeds of up to 100mph (the limit of my experience) the cars feel extremely safe and stable.

TYPE 50B

Further development of the Type 50 engine took place whilst Type 57 production was in full swing. Titled 50B, these highly-tuned engines were seen in various special racing cars in various forms and sizes from 3- to 4.7-litre. The latter size unit was fitted in a monoposto car (chassis number 50180) which also featured enormous hydraulic brakes and a four-speed gearbox which, I understand, may have incorporated a synchromesh mechanism. This car was certainly the fastest Bugatti ever built and achieved a certain amount of success, being particularly notable for winning the first postwar race held in the Bois-de-Boulogne in 1945.

The 3-litre version was a complete redesign involving a bolted-together cylinder block of aluminium but embodying bronze combustion chambers for each cylinder. This set-up was being tried by other racing car engineers who were interested in improving thermal efficiency, but the Bugatti version did not meet with success, possibly as a result of sealing problems.

Power output must have been satisfactory for we know that the 3-litre version which ran in the Cork Grand Prix was timed at 147mph during the race when running on seven cylinders! Other - stillborn - manifestations of the Type 50B concerned a scheme for a road speed record car which utilised two 4.7-litre engines coupled together. An aeroplane for speed record purposes, also fitted with two of these engines (one with left-hand rotation), was actually built, although it never flew. All the problems of 1939 signalled the end of further development.

LIVING WITH A TYPE 50

Ian Preston reminisces about his years of Bugatti Type 50 ownership. (Chassis no. 50139, registration LPG 61).

"I claim, with Peter Stubberfield, to have had the most extensive postwar motoring in a Type 50. I bought LPG 61 from 'Tom' Thomas in the late '60s and I don't think he did many miles in it. Before that it belonged to Peter Stubberfield and I remember it at Prescott with the single-seater T35B trailered behind. Peter has told me of epic journeys to hillclimbs, for instance, to Bo'Ness - a fair haul from his home near Maidenhead - and he also used it for towing caravans as part of his business. At that time, because it had the two spare wheels on side mounts slotted into the front wings, the tow bar was reasonably easy to fix and use. Although I achieved one of my ambitions with the car - to tow HED 21 to some hillclimbs and sprints - I always felt that the spare wheels were not aesthetically satisfactory in this position and made the car look too short. Also it was very light on the back end and rear wheelspin wasted some of the performance this fantastic engine could produce.

"During preparation for repainting

the car in about 1970, the slot in the bodywork and wood framing at the back of the boot, and areas of bare chassis paint on the back crossmember, satisfied me that the two spare wheels had originally been mounted on the back. The front suspension had also been 'improved' by the fitting of telescopic shock absorbers, which necessitated ugly bulges that spoilt the line of the inside of the front wings. In a box of bits that came with the car was a pair of tele-controlled Hartfords which, after a complete overhaul, were refitted. However, the hand wheel adjusting unit which went up to the dashboard was missing, so a local adjustment system was evolved. The remote control adjustment system was obtained but we never got around to installing this because we had been told that they were an unnecessary distraction, and if you did fiddle about with this mechanism the shock absorbers were never right for the next bump you hit.

"In discussion with Jack Lemon Burton on this matter he offered me a pair of De Rams, plus a dismantled unit for spares, advising that these were fitted onto the later delivered Types 50 and 46. With the help of Ronnie Symondson's article on restoring De Rams in *Bugantics* the units were overhauled, firstly making all the special tools. We were rather proud of ourselves when they were nicely fitted and, once we had mastered the proper adjustment, they were never touched again.

"Both these alterations back to original specification - one at the back and one at the front of the car - improved aesthetics and handling.

"The dreaded signs of water in the oil required investigation, which culminated in overhauling the engine. There was a horizontal crack at the bottom of the water jacket just below the top of the crankcase and, on further investigation, it was discovered that the side of the water jacket had been pushed out and disconnected from the outside of the cylinder bores. The reason for this was that the bottom of the water jacket, being below the top of the crankcase, could not be drained for winter/frost protection. This is pertinent to an oft-quoted letter on Ettore's autocracy when a new owner wrote to him complaining that one couldn't satisfactorily drain the block. Ettore's reply was that if the owner could afford one of his motor cars he should be able to afford a heated motor house to keep it in! Another foible of the design was that if you wanted to grind-in the valves on a T50 you had first to take out the back axle - but more of that later.

"The engine was initially rebuilt at Bassett Down Engineering by Nigel Arnold Forster and the only problem we had then was finding the right pistons. Nigel noticed a spare piston from Delage II looked about the right size, which it proved to be, so piston blanks were available. Luckily, the valves, etc., were in good condition.

"After the overhaul we really began to appreciate the performance of the Type 50. However, the Delage piston design obviously wasn't totally satisfactory because, on the way to a Prescott Meeting, the head came off piston number seven. By this time I was being helped by a jig and toolmaker, John Beech, and Nigel was busy so we decided to set about it and attempt the second engine rebuild ourselves. Although a rod was bent there were two spares, with the bits and pieces, new pistons were made using the original Bugatti drawings and, by this time, and having bent two valves, we decided to have a new set. These were machined locally, as were the pistons, by Silicom Pistons, and made from blanks supplied by a very helpful valve maker in Sheffield who usually made them for lorries.

"This is when we were made aware of the attention to detail of the design of this engine. The block casting for a twin cam, no separate cylinder head, extended between the bores to carry the nine main bearings for the crankshaft is, even by today's standards, a masterpiece. Everything fits onto the block: consequently, when it's all set up, it's there for good and there are no separate cylinder head gaskets or any other bits and pieces. The crankcase/sump is another piece of Bugatti's pattern making and casting art; it's a beautiful aluminium bath with sealed sections around the top, closing it onto the block. It is so light as all it needs to do is stop the oil dropping onto the road. However, some of the details of these seals are a bit fiddly to assemble, especially the Bugatti small section square rubber seals (were these, possibly, the first 'O' rings?).

"Although the car's performance had been spectacular, it had not been up to what Bugatti had claimed and some of the books (and Stafford East) had quoted. We did find the valve/camshaft timing was slightly out so, leaning on Hugh Conway for confident advice, we reset it.

"Yes, we did have to take the back axle out before we got the engine out! This bit of folklore has a basis in fact as on the T50 (and, of course, also the T46) Bugatti employed his patented rubber-mounted flywheel (luckily on our rebuilds this didn't need any attention as this, no doubt, could have been a problem) and in order to stop everything wobbling about there is another bearing at the back of the clutch shaft mounted in another Bugatti standard cast aluminium clutch cover, which bolts onto a substantial tubular crossmember. At the back of this is the normal flexible disc propshaft connector which, in order again to keep it in line because of the considerable torque from the engine (remembering that the gearbox is in the back axle), has a ball socket fitted into the centre of the spider. This needs to be moved back before anything can be moved upwards, as the engine will not go forwards enough to release it.

"We experienced another maintenance hassle involving the clutch which is of basic Bugatti multi-plate design comprising steel driven plates and, being a dry clutch, Ferodo-faced steel driving plates which transmit the load from serrated grooves on the inside of the flywheel. At between two and three thousand miles one finds the clutch becomes reluctant to free properly, a problem caused by the Ferodo dust being centrifugally compacted into these grooves. The first time we undertook rectification of this problem we were again faced with moving the rear axle back to take out the clutch assembly.

"Amongst the spares that came with the car were some Ferodo discs with serrated teeth around the periphery, which were obviously intended to replace the steel and Ferodo plates. At the time of this first clutch rebuild we wondered whether these were a good 'modification'. Luckily, before trying them we called in on Jack Lemon Burton on the Edgware Road, who had seen some of these plates before and, with his usual chuckle, advised that they were definitely not the thing, as the Ferodo dust sticking in the grooves had got even less space and consequently the clutch would seize in less than one thousand miles!

"Jack told us the way he used to clean out clutches on Types 46 and 50 was by undoing some of the bolts on the cross tube, pushing it about half-an-inch back and the clutch could then easily be dismantled. As he pointed out, it was all very well for the Bugatti agents to charge relatively high service charges. As it is today, if you take your modern motor to the main agent compared with your competent local garage man.

"Another modification Jack did was to drill a series of holes in the periphery of the flywheel at every 5th or 6th groove so that if, about every 2000 miles, one was to blast the compressed air gun into the clutch through these holes, quite a lot of the Ferodo dust would be blown out. Actually, the next time we did this we noted our flywheel was already drilled because, of course, the car had passed through Jack's hands before the war. Of course, today this clutch cleaning operation would have to be carried out by four men in white coats and special breathing equipment, with an environmental health certificate!

"Back to the rebuild. The revised valve timing proved a success and produced such an improvement that we can now vouch for Bugatti's claims of 110mph! On one occasion we had opened up along a reasonably straight section of dual carriageway not far from home and she was really flying, when suddenly I heard an awful noise. Thinking 'another rebuild,' I eased off quickly, to be immediately overtaken by a two-wheeled Japanese whirling wonder - phew, what a relief! At the end of the dual carriageway the motorcyclist slowed down so we stopped for a chat. I apologised for suddenly slowing down in front of him and, on explaining my reason, he commented that we were 'well over the ton,' so he agreed to trail us back down the other carriageway and back again for a definitive test. Our speedo was reading 180kph plus on both runs and the friendly motorcyclist confirmed that our speed had been over 110mph, even tickling into 112mph. This was very satisfying.

"The steering was superb. Admittedly, it was a bit of a haul in a limited parking space, but as long as you used the old lorry driver's technique of having some rolling movement it was not too bad, and once on the move it was much lighter than any other

This picture accompanied Hugh Conway's text in Bugantics.

comparable-sized and aged car I have ever driven. Long runs were enjoyable and it went round corners as though on rails.

"The brakes, once set up properly (with the front leading shoe linings tapered on plan by a third in length to avoid brake grab) were excellent, especially considering the size and weight of the car and must have been above average in the car's day.

"The gearbox, being in the back axle, is operated by quite long rods and, of course, there are only three forward gears (all that is needed) so that reverse is in the top of the spare gate. We did break a selector, but this coincided with the first time we had experienced the clutch sticking problem. However, Bill Bloomfield had a collection of spare back axle bits so this was soon repaired. Although the whole axle assembly with gearbox is rather a lump, the fact that it all splits along its vertical axis makes it easy to lay the assembly in and drop the other half back into position: we did this by placing the lower half of the gearbox in the workshop pit.

"Although I've recounted some of the problems we had with LPG 61 over about ten years that we owned it, once sorted out we did a fair amount of trouble-free mileage - about 3000 miles a year - going on several rallies, including those on the Continent, competing in Bugatti handicap at Prescott, and for general use as a second family car. We only decided to part with the car because of the opportunity of thereby getting a Type 43 and the fact that as I was the only person using a Type 50 then, the thought of a new cylinder block, should it ever need one, was frightening. Of course, now new blocks have been made and I remember seeing three being machined at Mr. Lawrence's works a few years ago.

"It had been an exhilarating and gratifying experience to have owned and used one of these fantastic examples of Ettore Bugatti's motor cars."

THE THIRD LE MANS CAR

The following text by the late Hugh Conway first appeared in *Bugantics* magazine under the heading "Le Mans Type 50?" in Spring 1988 (Vol 51, No.1).

"One of the more interesting cars referred to in the original Bugatti Book was a Type 50 stated to be a Le Mans 4-seater, and certainly looking very like the cars which ran in 1931, and withdrawn by the factory when one crashed from tyre trouble during the race. The chassis [number] was given as 50143 and engine [number] as 147, but as we started to revise the Register which was based on factory records we ran into trouble. These records give the car as being delivered to Antoine Schuman of Paris in December 1931, which was after the race at Le Mans, and he was, we thought, a wealthy banker unlikely to want such a sporting type.

"However, Jack Lemon Burton had imported the car in April 1937, selling it the next month to Jack Crowther; the registration then was DAR 349, which was issued in about May or June 1937. (It was later re-registered with the lower tax in 1950 as NPP 128, August 1950).

"Pam Crowther now picks up the story: 'Before he let Jack have it he took him out in it and said "Remember it is twice as large, twice as fast and with half the stopping power - therefore, focus much further ahead than you normally do." We took it to the Brighton speed trials (against Craig who had the same engine put in a Type 55); he won the first trip, owing to our wheelslip, so Jack put back the spare wheel and me, among other things, and we won the second trip - just didn't touch 100mph in the quarter mile. All we did to the car was to arrange a wraparound windscreen for comfort - unscrewing, so as to replace the racing screens as required. It had a 40 gallon tank and did 8mpg touring.

" 'Came the war - it would have deteriorated in a damp shed - so we sold it for £500 to K. W. Bear who could keep it decently, agreeing between friends, if he wanted to sell, we might have first refusal. He lowered the body several inches (silly ass!) but, of course, when he crashed and was killed at Jersey, the car went to Stafford East.'

"We now turn to Miles Coverdale who had, by now, what seemed a genuine 1931 works car. 'When I was finally able, after some fifteen years, to persuade my brother to sell me the 1931 Type 50 Le Mans Bugatti that he owned, I began at once to try and identify which of the two surviving cars I actually had and where it had been in the intervening years. In the process I did turn up the story and origin of the Type 50 Le Mans car presently in England, familiar at Prescott after the war as DAR 349 in the hands of the late K. W. Bear.

" 'It was Robert Jarraud, author of the book *Bugatti Double Arbres*, who

K. W. Bear charges along in DAR 349, the ex-Le Mans car.

brought to my attention the phantom Type 50 which was missing from the start of the 1935 Le Mans race, car no. 3, which reportedly had belonged to Richer Delavau but been wrecked in practice by his co-driver and cousin Jacques de Valence. Jarraud kindly undertook to contact Delavau for further information on the car to see where it might fit into the story, but results were not too satisfactory and the car could not be identified from the information provided. However, on a visit to France to the home of 'Le Grand Robert' Aumaitre, Bugatti factory mechanic with the Type 50 cars at Le Mans in 1931, 1933, 1934 and 1935, I was given the name and address of Delavau's mechanic for the 1935 attempt and so I had another opportunity to identify this car.

"'Paul Delaitre responded enthusiastically to my appeal for information about the missing 1935 entry and, most importantly, was able to send me two photographs of the car, one taken before and one after the accident. From these it can be seen without question that it was, in fact, this car, completely rebuilt after the accident, which was sold in England in the late 1930s as a 'Le Mans Type 50.'

"The story from Delaitre is as follows: 'Delavau bought a Type 50 Coupé with Figoni body, then removed the body to be sold back to Figoni. From the Bugatti factory he obtained a used four-seater Le Mans-type body (possibly from the practice car, or the missing survivor from the 1931 race), and mounted it on his chassis. Then, with various changes to the chassis, supercharger, gear ratios, etc., and with tuning supervised by Maurice Rost who drove in 1931, the car was ready to race when it was crashed in the last night-time practice.'

"This, then, is the story of DAR 349, which, although not part of the factory team in 1931, and although it never got to race in 1935, can still be considered a Le Mans 50."

BUGATTI

46/50

II

FIXED HEAD COUPÉS & 2-DOOR SALOONS

Main picture: Chiron taking delivery of a car (probably 50147) at Molsheim. This same Type 50 Fixed Head Coupé of 1930-31 is also pictured below: it was almost certainly built at Molsheim. Note the vertical lines, ubiquitous pram irons on the rear quarters and exposed wheel nuts.
Inset: A Type 46 factory Fixed Head Coupé similar to the Chiron Type 50, but with wire wheels covered with discs.

This page: A neat Type 46 Fixed Head Coupé with fabric top and false pram irons. This is a typical example of 1930-31 styling and the work of Gaston Grummer of Clichy.

A late Type 46 or 50 Fixed Head Coupé, possibly by Chapron, outside the Lemon Burton residence in North London upon arrival from Paris in the mid-to-late thirties. Note the Cleveland petrol sign showing the price of petrol as seven pence a gallon in 'new' money!

Another Type 46 which was later fitted with a Type 50 engine. Chassis number unknown.

A late 46 or 50 with "Faux Cabriolet" body, photographed in France circa 1960. Coachbuilder unknown

Below & opposite page, top: Another Fixed Head by Gaston Grummer, clearly a concours winner typifying 1930s glamour. Two stars at Longchamps. Chassis no. 46405.

Below: Type 50, chassis no. 50146. Rebodied to a high standard in the fifties (by Ghia?)

AGO 37

Left and opposite page: The penultimate Type 46S, chassis no. 46587. Sold by the London Depôt, fitted with a Fixed Head Coupé body by James Young and registered in July 1933. It's still in perfect original condition today.

Below & inset: Type 50. A factory-designed and built Fixed Head Coupé (note Bugatti patent door handle). The American, mainly Chrysler, influence, allied to Bugatti flair, results in outstanding elegance.

Type 46. Two very similar 2-door Saloons (or Coaches). The car with the rear-mounted spare wheels is considered to be the work of Van Vooren.

Type 46. The standard Bugatti-designed coach built by various sub-contractors. This example, with the then popular composite finish - metal panelled below the waistline and fabric above - is credited to Visse et Haf.

Below and next page: Type 46 (chassis no. 46219) to a Bugatti design built by Weymann of Paris. This car, sold by the Brixton Road Depôt, was driven to the Arctic Circle by Bunty Scott-Moncrieff in the postwar years where it is pictured astride the Arctic Circle. Eric Giles, later president of the Bugatti Owners' Club, used the Scott-Moncrieff Type 46 for fire-fighting duties during the war. The photograph shows the car with headlamp masks and regulation white painted edges to wings.

GC 8581

Type 46S (chassis no. 46585) by Billeter & Cartier of Lyons on the left, together with chassis no. 46241, a Bugatti Coach built by Gangloff.

A recent photograph of 46585S.

Type 46. Standard Coach, most probably Gangloff-built, photographed at a Paris salon.

Left and below: Type 46. Sold through the London Depôt, a standard 2-door composite Coach by Weymann of Paris. Chassis no. 46163.

Type 46. Colonel Sorel, manager of the British Bugatti Depôt, opens the course at an early Bugatti Owners' Club hillclimb. Chassis no. 46219.

Type 46, chassis no. 46241. A late standard Coach by Gangloff recently converted to alloy wheels.

Type 46. A late standard Coach by Weymann of Paris.

Type 46. The well-known Fiacre style built at Molsheim to the design of Ettore. Note the well-tended garden of the Bugatti residence.

Opposite, top: Another Fixed Head Coupé styled by Bugatti, but probably built by Gangloff.

Opposite, bottom: A cumbersome Fixed Head Coupé photographed later in life, coachbuilder unknown.

Above and below: Type 50. A later version of the Fiacre. In this case, the design of Walter, the resident Bugatti body designer. In the lower picture the car is photographed in front of the factory doors, a favourite backdrop.

Chassis no. 46540S, an unusual and stylish coupé in Sweden. Coachwork by Jos Neuss, Berlin.

Type 50. A striking Fiacre by Bugatti. Note the unusual hub treatment. This car was owned by the Marquis of Cholmondeley. Note Jean Bugatti wing line.

This page: The first Semi-Profilée body. The windscreen wipers were pivoted at the bottom of the windscreen, which may have been more effective than the original top mounting in keeping wiper pressure on the screen. Wimille looks pleased with his charge!

Both pages: Type 46, chassis no. 46546. Semi-Profilée Coach believed to be the only Type 46 originally fitted with a body of this type. The car is now in the USA.

Type 50. A Coach, Semi-Profilée, styled by Jean Bugatti and built at Molsheim. This car, chassis no. 50169, was delivered to Armandias, the head of the Messier undercarriage firm (much later to join with Bugatti under Hispano supervision). The car is here seen with its second owner. M. Lesoure, while he was engaged in the French equivalent of the Territorial Army.

Type 50. Front end view shows the Jean Bugatti flutes on the front wing centre line of this Semi-Profilée-styled car.

Oppposite top and this page, below: Type 46 Profilée. Works photographs of the interior, which doesn't have a rev counter. Note the ornate door trim pattern.

Type 50. Bugatti drivers Veyron, Dreyfus (passenger seat) and Wimille (running board) photographed in the works Type 50 Semi-Profilée in front of the Bugatti dovecote at Molsheim.

A Semi-Profilée Type 50 in a period setting. Trilby hats seem popular!

Type 50. The first Full Profilée body on a Type 50, built at Molsheim in 1932.

Front view of the first Full Profilée body on a Type 50, built at Molsheim in 1932 ...

... a front bumper was fitted in time for the Paris Salon.

Left and centre left: Type 46. Rebodied in Coach Profilée style by Andy Rheault in the USA. Now converted to alloy wheels and in the Schlumpf Museum. Chassis no. 46482.

Above & below:Type 50, chassis no. 50117. This 3-passenger Coupé by Million Guiet, owned by Ken Purdy, was 1st prize and blue ribbon winner in the Classic category at the 1961 New York Auto show.

Type 50. Short chassis, 2-door Saloon by Million-Guiet. This body was heavily modified in the forties, but is now being restored to original state.

Type 50. A Gangloff 2-door Coach at the 1931 Paris Salon.

BUGATTI

46/50

III

4-DOOR SALOONS

Type 46; believed to be the first 7-seater and probably chassis no. 46135.

Right & opposite page: Another 7-seater which was used as a demonstrator, believed to be chassis no. 46136. In this photo on the right, young Roland Bugatti sits on the running board and the attractive lady is thought to be his English governess.

Type 46. An early fabric Saloon, probably by Freestone & Webb of London, built under Weymann licence. This car featured in the British Depôt catalogue and was probably chassis no. 46184 or 46209.

Type 46. An early Four Light Composite Saloon body by Gallé of Paris. It features wire wheels embellished with discs and typical Art Deco front bumper.

Type 46. 4-door, Close-Coupled Saloon with fabric top and lavish chromium or stainless belt line mouldings by Van den Plas of Brussels.

This page: Type 46. Two versions of a later version of the standard full Saloon, Bugatti-designed but probably built by Gangloff. It is likely that one or two examples were fitted with a partition between the chauffeur and rear passengers. Certainly, these cars often incorporated two occasional seats, thus providing space for seven people. One car has Rudge wire, but with easy-clean discs, and the other has aluminium wheels. Photographed at Molsheim.

Type 46. A standard Coach imported by the Brixton Road Depôt and also featuring in the catalogue. Note the English chauffeur and suitable background. Probably chassis no. 46163.

BUGATTI

Type 46. This car was sold by Brixton Road Bugatti but fitted with a Sunbeam Saloon body, probably supplied by Cooper Motor Bodies of Putney, who specialized in the purchase and resale of 'end of run' and surplus bodies from manufacturers. Fitted to the chassis by P. G. Twigg of Teddington. Chassis no. 46257.

Chassis no. 46531 photographed at Prescott in 1938 after Jack Lemon Burton bought the car from Col. Wilfred Sorel, London Depot manager. The car has the late addition of radiator slats, similar to a Type 57, a change of colour, and has been registered with a 1938 London registration number, presumably to disguise its age! Rose Lemon Burton is standing by the car.

A four light pillarless Saloon by Gangloff, approximately 1931. The pillarless style was not a very good idea but, nevertheless, enjoyed a vogue for several years.

Type 46. A late example - with alloy wheels and hub caps - of the work of Freestone & Webb, chassis no. 46533.

Type 46, chassis no. 46531. The personal car of Colonel Sorel, General Manager of the London Depôt, this is a Close-Coupled Sports Saloon by Arthur Mulliner of Northampton. The car is pictured opening the course at a Bugatti Owners' Club hillclimb and in London's Hyde Park (Bayswater Lodge).

Type 46 or 50. Sedanca de Ville by a Parisian coachbuilder (may be Kellner, judging by the door handles).

Left & below: Type 46. A late Sedanca de Ville at a Concours D'Elegance held at the Longchamps race course, considered to be the work of Kellner.

Type 46. Rebuilt and modified postwar by Georges Combe of Paris, this car was equipped for a journey across the Sahara Desert. The trip, however, was not undertaken. Chassis no. 46331.

Opposite: Two Type 50s, the work of Million-Guiet. These bodies were of all-steel construction and bolted together under Vizcaya patent.

Chassis no. 46331, Figoni Drop Head Coupé. This original body is now reunited with the chassis and is owned by Lord Raglan.

Chassis no. 46355, Van Vooren Drop Head Coupé. This photograph was taken outside the Pélé Hall Hotel, North Wales, in 1987.

MILLION-GUIET

A stylish close-coupled full Saloon by Gangloff which could be a late Type 46 or Type 50, which was delivered to an owner in Tunis.

Opposite page: Type 46. Chassis no. 46510. A 46S which remained unused for many years and was later fitted with a Gangloff Saloon body of much later date than the chassis. Now in the Schlumpf collection.

A bizarre Type 50 4-door Saloon, coachbuilder unknown.

A very late Type 46. chassis no. 46580, with a pillarless saloon by Van Vooren. A youthful Tony Rippon, now President of the Bugatti Owners' Club, stands with the stopwatch.

Ettore Bugatti welcomes President Le Brun to his stand at the Paris Salon in 1933. A Type 55 stands behind.

Type 46. A late car with coachwork by Lancefield of North London.

The third Lancefield body on chassis no. 46535. This picture was taken at Prescott circa 1964.

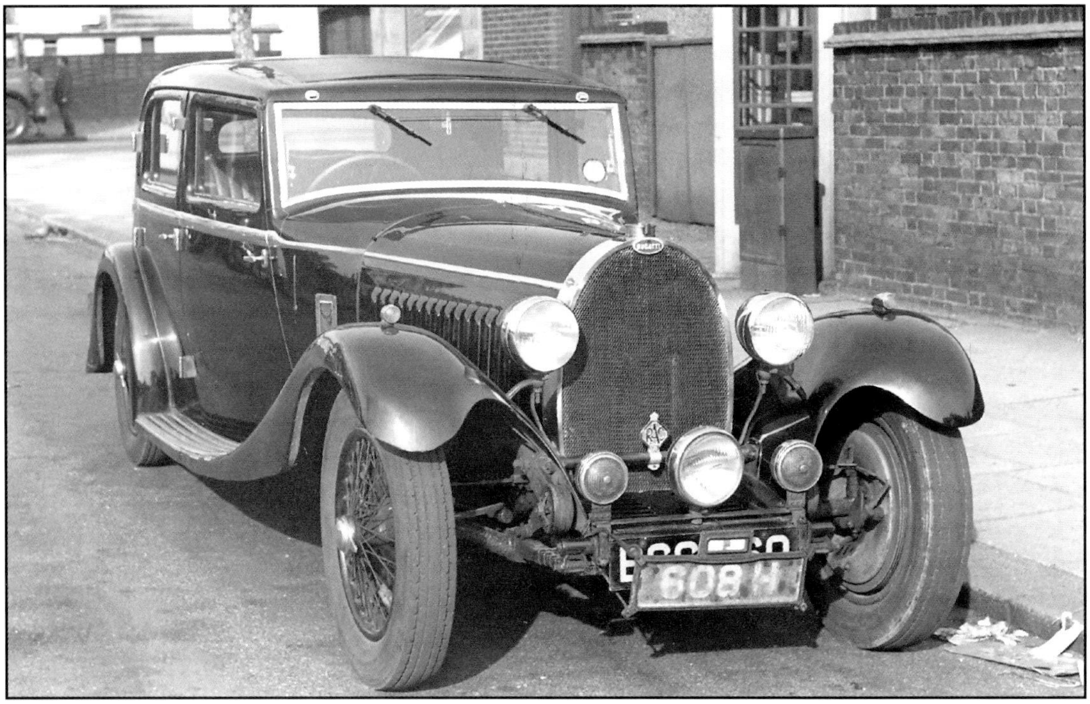

This page and opposite, top: Another, more angular, Lancefield saloon, chassis number 46534, when in the ownership of Jack Lemon Burton, and prior to sale to Sam Clutton, who owned it for many years.

Chassis no. 46577, sports saloon by Lancefield of North London, later owned by Eric Giles.

IV

CONVERTIBLES

Type 46. Bizarre four-seater sports body by Corsica of Edgware Road. This car has disappeared without trace, as have so many of the 'English' cars, a result of the horsepower tax and corresponding unsaleability of such cars on the secondhand market. Many were scrapped in the period just before and during the early years of the war. The style is typically Art Deco.

Type 46. Drop Head Coupé by Van Vooren. This was an early car, originally owned by R. B. Pope, a well-known Bugatti enthusiast.

Type 46. An attractive Drop Head Coupé credited to the little-known firm of Visse et Haf. Note the multi-louvred bonnet and valances, cycle wings and lack of running boards.

Type 46, chassis no. 46501. A standard Roadster with a conventional wing line.

Type 46, chassis no. 46355. A Van Vooren Drop Head Coupé, fitted (recently) with alloy wheels, which has successfully completed the Peking-Paris marathon.

Type 46. Gangloff Drop Head Coupé which was later (postwar) fitted with a 50S engine (ex-chassis no. 50123) in the USA.

Type 46, chassis no. 46287. Rebuilt in the USA with this hideous special body utilising parts of a Cadillac. Now in the Schlumpf collection.

Type 46, chassis no. 46340. A well-known car rebuilt with a canework body by a Dutch school for the blind.

This page: A late Type 46 Roadster with forward opening doors, golf club locker and double swivelling windscreen. Built by the obscure coachbuilder, Ottin. Chassis no. 46501.

Both pages: Type 46, chassis no. 46571. A late Drop Head Coupé by Brainsby-Woollard at Prescott in the early fifties (top left) and the same car, now in the USA.

This page: The last Type 46, chassis no. 46588. A James Young Drop Head Coupé, eventually registered in London in 1935. Note the modern postwar Standard Vanguard front bumper and radiator slats.

A Type 46 rebodied in the late thirties, coachbuilder unknown. Chassis no. 46456.

An unusual Cabriolet, bearing the stamp of the English firm of Brainsby-Wollard who specialized in this type of coachwork.

A Type 50 factory-built Roadster. US influence is evident in the style, which was also used - suitably scaled down - on Types 40A and 43A.

Right & above right: Two pictures of an identical body on a Type 46 - note the wire wheels.

1767
NV4

Both pages: Type 50. The striking, factory-built Coupé incorporating a 'Landaulet' top. This style is credited to Jean Bugatti, then only 22 years old. The car pictured by the Molsheim factory door is probably the first of this model (chassis no. 50112). Note the bronze 'Bugatti'-designed hinges on the factory door.

1244-W5

Sir Robert Bird's Type 50, chassis number 50118 . Ready for despatch at Molsheim in front of the Rembrandt Bugatti Museum; head fixed ...

... and head lowered.

... the same car, some time later, at a Bugatti Owners' Club hillclimb held on a building site at Chalfont St Giles circa 1934.

Type 50. The catalogue Drop Head Coupé, Bugatti-designed and executed by Gangloff Fréres of Colmar. Always the principal sub-contractor to Bugatti, Gangloff specialized, in particular, in Drop Head Coupés, together with special orders and modifications to suit owners' wishes.

PZ-82-50

Opposite page: Type 50, chassis no. 50128. Recent photographs of a very original car with a body by Sodomka of Prague - the car is still there!

This page: A nearly identical car in the USA.

Opposite page: Jack Lemon Burton imported this car from Paris in the late thirties. The Bugatti-built coachwork is similar to that of a Type 43A. These pictures were taken at Brooklands.

Drop Head Coupé on a Type 46 by Graber.

This page: Type 50, chassis no. 50139. This Drop Head Coupé, to Bugatti catalogue design by Gangloff, is the Stubberfield/ Preston car described earlier in this book.

Type 50. The coachbuilder of this Drop Head Coupé was Weinberger of Munich.

An unusual Roadster-Coupé. Note the unusual bonnet/scuttle line and fancy paint pattern around the hubs. The coachbuilder is unknown.

This Drop Head Coupé features a well-raked windscreen. Probably Gangloff.

Works driver Meo Costantini with a Type 50 Coupé.

Type 50. A Drop Head Coupé on French licence plates but bearing British BRDC and BARC badges. Considered to be the car of Sir Robert and Lady Bird.

This page: Type 50. Cockpit views of the Roadster. Note the large rev counter.

This page and overleaf, top: The design of this Coupé appears to be later than the chassis. Valanced wings indicate 1935 or later.

Chassis no. 46136 rebodied in hideous fashion - it looks eastern European.

A rebodied Type 46 considered to be the work of Sodomka of Prague.

Type 50. A unique radiator shell combining slats on a late chassis.

This unknown car received a serious "scrape" along the nearside near Bordeaux.

Type 50, chassis no. 50126. Rebodied postwar by Abbots of Farnham, England. Repainted in traditional Bugatti-style (inset).

DS 210

BUGATTI 46/50

V

MOTORSPORT

The Chiron/Varzi (4) car in the 1931 Le Mans 24 Hours race: an event that was ill-starred for Bugatti.

The Divo/Bouriat car (5) corners hard in the 1931 Le Mans 24 Hour race before the Bugatti team was withdrawn.

This page and overleaf, top: The wreckage of the Conelli/Rost car which crashed out of the 1931 Le Mans race following tyre problems. The accident happened on lap 20 whilst Rost was driving and, though Rost was unhurt, tragically, a spectator was killed in the incident. In view of the tragedy, and because the tyres of the other two team cars were also deteriorating, the Bugatti team was withdrawn from the event. Rost had been in 2nd place when he lost control of the car and left the track.

Singleton Type 50s appeared again at Le Mans in 1933, 1934 and 1935 - retiring before the finish in each of these races. Here is the start of the 1934 event for which a number of Bugatti types were entered. The Labric/Veyron T50 retired on lap 73 when 16th.

The Labric/Veyron T50 in the 1935 Le Mans race. The car was in 2nd place when its transmission failed on lap 116.

Chassis no. 50143 prior to the 1935 Le Mans 24-Hour race. Unfortunately, the car was crashed during practice and never made it to the race ...

... but the same car surfaced in England in the late 1930s and, with its body lowered, became a familiar sight at the Prescott Hillclimb and other club events in the hands of K. W. Bear.

Above, right & opposite, top: More pictures of T50 50143, the ex-Le Mans car, in British club motorsport events ...

Below & overleaf, top: ... the car's licence plate number was later changed to NPP 128.

Opposite page: Type 50, chassis no. 50177. A surviving Le Mans car beautifully maintained by Miles Coverdale of Long Island, New York.

Overleaf: Type 46. The late Bill Bloomfield competing at Prescott in his 'Grand Prix'-bodied 46, chassis no. 46354.

5
CENTENAIRE BUGATTI
5
Crédit Mutuel
CMDP
HX-8415
NY HISTORICAL

5
HX-8415

GP 46

VI

ANATOMY OF THE TYPE 46 & 50 BUGATTIS

First Type 46 at Molsheim. This view clearly shows the intake side of the new engine, together with steering box and carburettor detail.

Another view of what is thought to be the original T46 at Molsheim.

Again, what is considered to be the original T46 at Molsheim. Note this early engine's lack of Bugatti nameplate lozenges.

T46 clutch housing and exhaust manifolding in close-up detail.

An original factory photograph of the first car, chassis no. 46120, showing front brake operation. Note Royale chassis in the background, perhaps indicating a small shop for special project assembly and experimental work.

Two more pictures of what is thought to be the original T46 under construction at Molsheim. Above is the combined gearbox/differential unit and, to the right, a view of the rear of the engine. Note the narrowness of the cylinder block.

BUGATTI

Induction side of a Type 46S engine; the right-angle supercharger drive (close-up on left) is similar to that of a Type 50. Note the square section blow-off valve above the supercharger.

Close-up of a rear engine bearer. Note the elegant sculptured bronze casting - typical Bugatti.

Opposite: A recently-restored Type 46 chassis.

This view of a T46 engine shows the polished heat shields fitted, as standard, over the exhaust manifolds.

Above: A recently-restored Type 46S chassis.

Interior view showing one of the floor tread plate castings; a typical feature found on several high grade French cars of the period.

One end of the front axle showing typical Bugatti tapered and polished steering levers and front spring box.

Axle and gearbox internals exposed.

The cast alloy front wheels designed for the Type 50 (but also used on the T46), axle, track rod, brake backplate and hub.

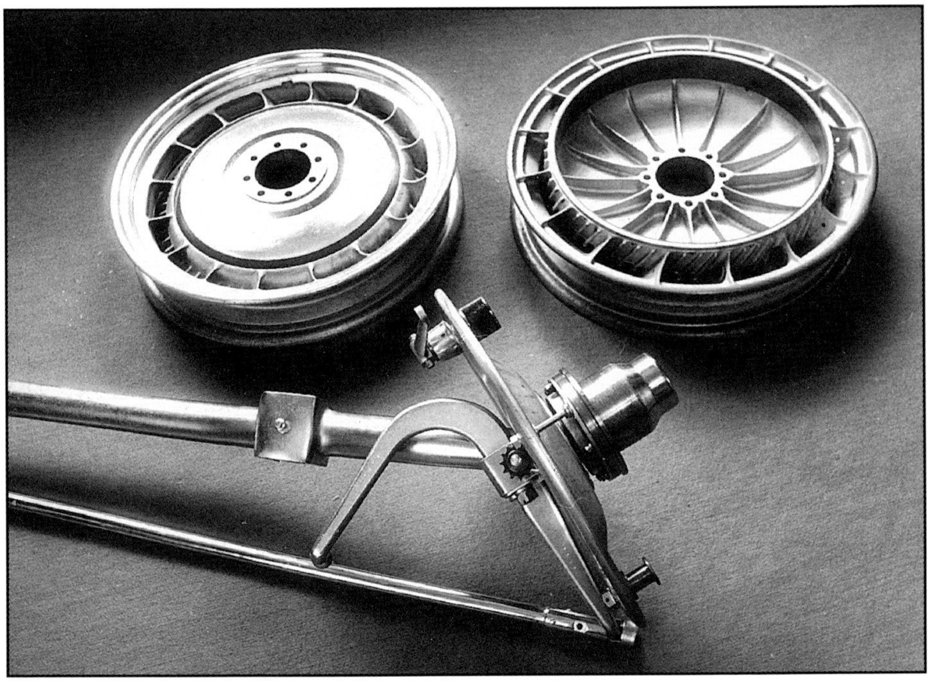

Front suspension detail showing André-Hartford tele-control shock absorber.

Steering box showing bolt-on drop arm common only to the'Big Bugattis.'

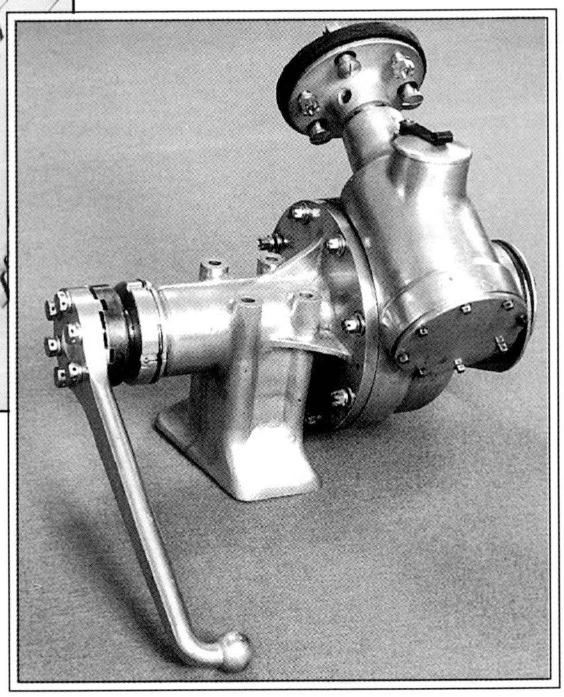

Detail of chassis in clutch housing area. Note dynamo drive and starter motor.

Right and below: Twin overhead camshaft Type 50 engine mounted in chassis and details of supercharger and carburettors.

The Type 50 twin overhead camshaft engine from the induction side.

The Type 50 twin overhead camshaft engine from the exhaust side (manifolds removed). The scroll-shaped unit toward the front is the water pump. Note the unusual uneveness of the turned surface finish on this unit.

Here, during rebuild, a T50 engine is upside-down as the crankshaft is lifted away. Note that the crankshaft 'webs' are circular and that the journals are balanced by pairs of circular discs.

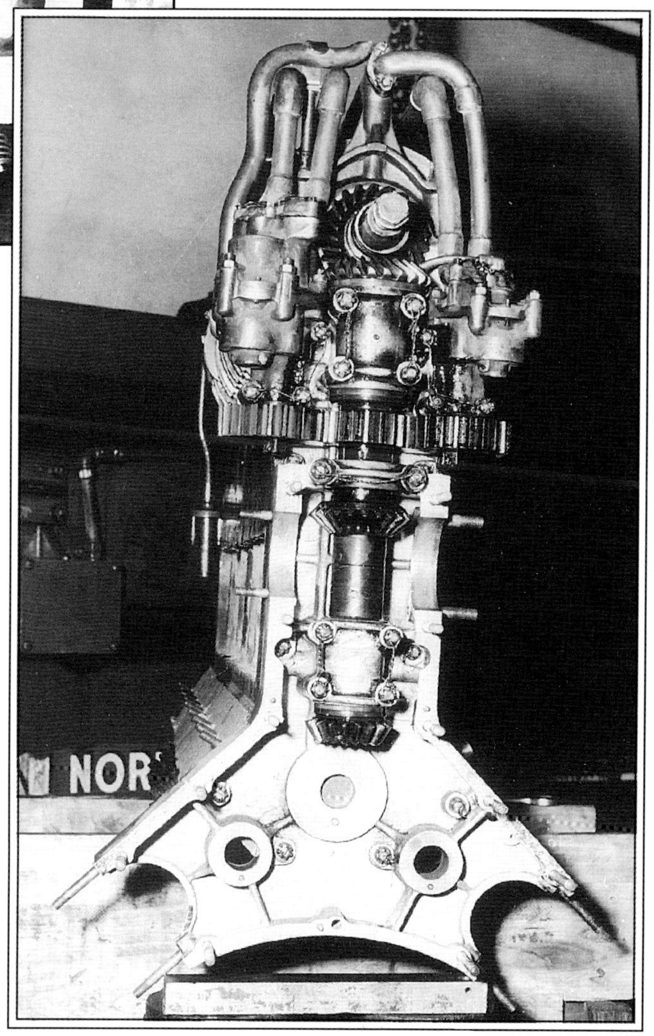

Again pictured during a rebuild, this T50 engine is upside-down. The units each side of the crankshaft bevel pinion are oil pumps with their associated pipework. The vertical shaft running between crankshaft and 'cylinder head' is the camshaft driveshaft: the bevel gear halfway up the shaft transmits drive to the water pump and supercharger on their respective sides of the cylinder block.

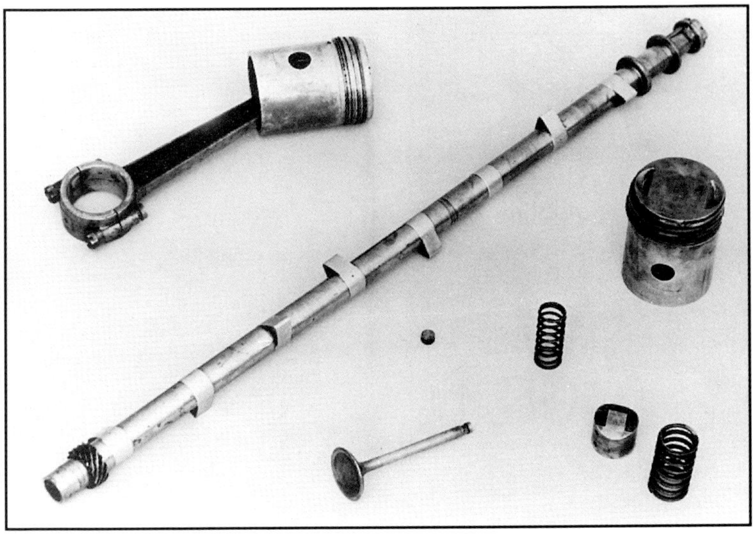

Some Type 50 engine components. Clockwise from top left - piston on connecting rod, camshaft, piston, valve springs and cam follower, valve.

Left: Front view of a Type 50 engine (minus supercharger). The fins at the base of the sump (oil pan) help to radiate heat and therefore cool the engine oil.

Photographed at the works, this is believed to be the first T50 chassis (no. 50112).

A small (155 x 220mm) Type 50 specification sheet as issued by Molsheim. The document is dated October 1st 1932 and is printed in black over a sepia coloured Bugatti 'watermark.'

CHASSIS TYPE 50 T

A COMPRESSEUR

CARACTÉRISTIQUES

MOTEUR à 8 cylindres en ligne monobloc.
ALÉSAGE: 86 mm. **COURSE**: 107 mm.
CYLINDRÉE: 4900 cc.
VILEBREQUIN à 9 paliers, supporté par le bloc cylindre.
2 ARBRES A CAMES en tête.
2 SOUPAPES PAR CYLINDRE.
2 CARBURATEURS.
COMPRESSEUR toujours en prise commandé par le moteur.
ALLUMAGE par batterie, bobine et distributeur.
EMBRAYAGE à disques multiples fonctionnant à sec.
BOITE DE VITESSES placée dans le carter central du pont AR.
ESSIEU AV creux en acier à haute résistance.
DIRECTION à droite.
ROUES aluminium coulé brevet Bugatti.
PNEUMATIQUES de 32 × 6,00.
FREINS sur les 4 roues, avec levier à servo-régulateur.
SUSPENSION: Ressorts 1/2 elliptiques à l'AV - 1/4 elliptiques à l'AR (brevet Bugatti).
ÉCLAIRAGE ET DÉMARRAGE électriques par éléments séparés.
EMPATTEMENT: 3,50 m.
ENTRÉE DE CARROSSERIE: 1,95 m.
EMPLACEMENT DE CARROSSERIE: 2,60 m.
VOIES AV et AR: 1,40 m.
POIDS DU CHASSIS NU A VIDE: Environ 1300 kg.

PLAN DU CHASSIS EN ÉLÉVATION

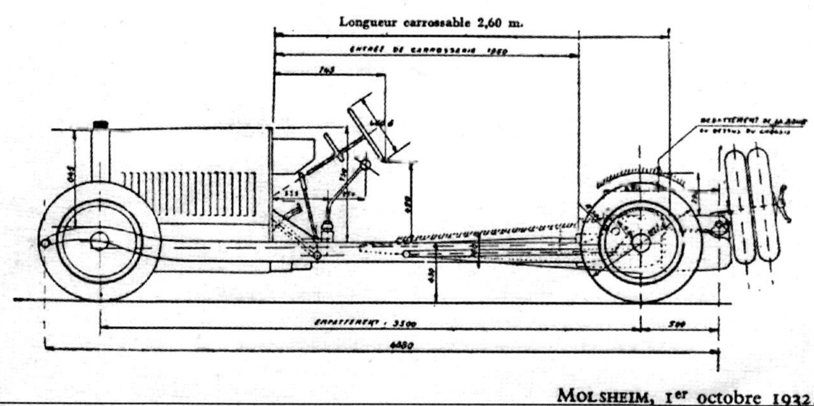

MOLSHEIM, 1er octobre 1932.

This page: Front and back of a single sheet brochure for the Type 46 as issued by Bugatti's London Depôt. The document is printed on parchment type paper in red and black. Size is 216 x 282mm. Date of issue unknown but believed to be 1930.

8-CYLINDER
5-LITRE TYPE 46

4/5 SEATER "SPORTSMAN" SALOON MODEL

(Particulars of other types of coachwork supplied on request).

THE 5-litre model is offered with every confidence as certain to fulfil the requirements of the motorist seeking the true 'DUAL PURPOSE' car. Silence, both of the engine and transmission, coupled with exteme flexibility, makes it the ideal car for town usage. In catering for these requirements traditional BUGATTI characteristics have been retained, ensuring rapid acceleration, high road speed and safety for fast touring purposes.

BUGATTI

Chassis specification overleaf

CHASSIS SPECIFICATION . . .

Number of cylinders	8	Body space	9' 6"
Bore	81 mm	Overall dimensions	15' × 5' 7"
Stroke	130 mm	Wheels	Well base wire
Valves to each cylinder, operated by single overhead camshaft	3	Tyres	32 × 6.00
Number of engine bearings	9 Plain	Weight (approx.)	25 cwt.
Ignition	Dual high tension	Petrol tank capacity (approx.)	23 galls.
Type of Carburettor	Smith multiple jet	Petrol consumption (approx.)	12/14 m.p.g.
Gear ratios	1st 9.8—1 2nd 5.45—1 3rd 3.9—1	Oil consumption (approx.)	800/1000 m.p.g.
Back axle ratio	12 × 47	Minimum and maximum speeds on top gear	4/85 m.p.h. (approx.)
Wheel base	11' 6"	Treasury Rating	32.5 h.p.
Track	4' 7"	Tax	£24-15-0
Chassis price delivered London	Unsupercharged . . £975 Supercharged . . . £1175		

Conditions of Sale Guarantee and Warranty as published on the back of our official Chassis and Car Order Form.

ETTORE BUGATTI

1-3 BRIXTON ROAD LONDON S.W. 9.

TELEGRAMS:
BUGATTIMO CLAPROAD, LONDON
TELEPHONE: RELIANCE 3165-6

BUGATTI

8-CYLINDER 4.900 LITRE TYPE 50. SUPERCHARGED

TYPE 50 CHASSIS 'SPORTS' . . . . .

INTRODUCED to challenge the world's fastest standard touring or sports models, the 4,900 litre Type 50 BUGATTI opens a new chapter in the progress of motor travel. Whilst claiming this efficiency, this model chassis is quite light in the hand and consequently not tiring to drive

The brakes, steering and general road-worthiness are up to the highest BUGATTI standard, always aiming for safety. The 'Touring' model chassis is note-worthy for docility and flexibility. The 'Sports' model chassis should satisfy those especially seeking a very high road speed.

BUGATTI

Chassis specification overleaf

This page: Front and back of a single sheet brochure for the Type 50 as issued by Bugatti's London Depôt. The document is printed on parchment type paper in red and black. Size is 216 x 282mm. Date of issue unknown. Interestingly, the price section of the specification has been revised by pasting new prices over the originals.

BUGATTI

CHASSIS SPECIFICATION . . . .

	TOURING	SPORTS
Number of cylinders	8	8
Bore	86 mm.	86 mm.
Stroke . . .	107 mm.	107 mm.
Cylinder capacity .	4840 c.c.	4840 c.c.
Valves to each cylinder operated by twin overhead camshafts . .	2	2
Number of engine bearings . .	9 Plain	9 Plain
Ignition . . .	High tension	High tension
Type of carburettor	Two Scheblers	Two Scheblers
Gear ratios . .	1st 9.8—1 2nd 5.45—1 3rd 3.9—1	1st 8.6—1 2nd 4.8—1 3rd 3.4—1
Back axle ratio .	12 × 47	13 × 45
Wheelbase . .	11' 6"	10' 2"
Track . .	4' 7"	4' 7"

	TOURING	SPORTS
Body space . .	9' 6"	8' 1"
Overall dimensions	15'× 5'7"	14' 10" × 5' 7"
Wheels . . .	Cast aluminium well-base	Cast aluminium well-base
Tyres . . .	32 × 6.00	32 × 6.00
Weight (approx.) .	23 cwt.	23 cwt.
Petrol tank capacity (approx.) . .	23 galls.	23 galls.
Petrol consumption (approx.) . .	9/10 m.p.g.	9/10 m.p.g.
Oil consumption (approx.) . .	750/1000 m.p.g.	750/1000 m.p.g.
Minimum and maximum speeds on top gear . .	5/100 m.p.h.	5/112 m.p.h.
Treasury Rating .	36.6 h.p.	36.6 h.p.
Tax	£27 15 0	£27 15 0

Chassis price delivered London, Touring model £1900, Sports model £2000

Conditions of Sale Guarantee and Warranty as published on the back of our official Chassis and Car Order Form

ETTORE BUGATTI

1 - 3 BRIXTON ROAD
LONDON S.W.9.

TELEGRAMS:
BUGATTIMO, CLAPROAD, LONDON
TELEPHONE: RELIANCE 3165-6

This page: A catalogue of Bugatti coachwork for 1930 showing the bodies of Weymann of Addlestone, Surrey, England. The document measures 248 x 185mm and is produced in colour.

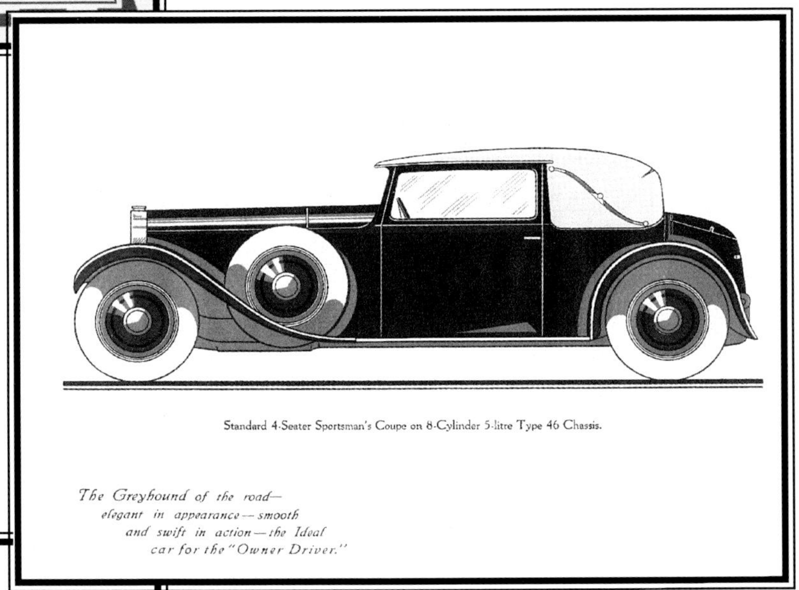

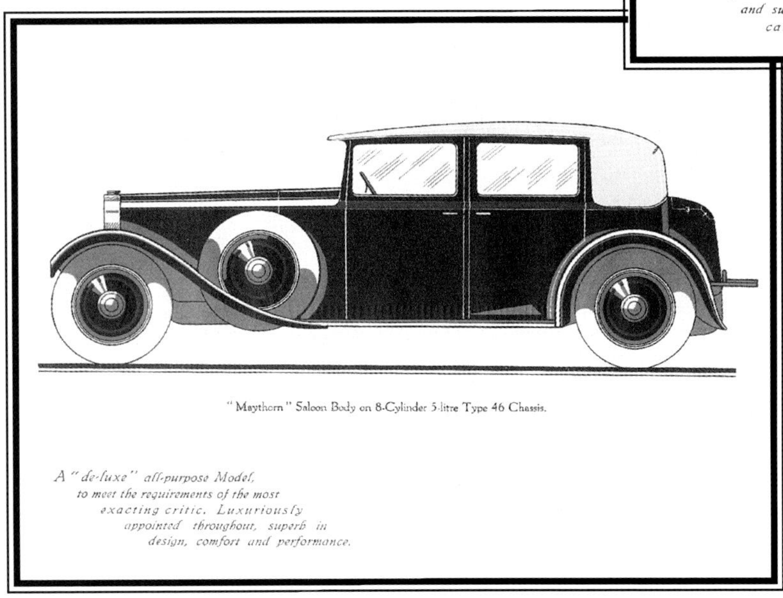

Following four pages: Reproduction of an original French-language Bugatti brochure which is not dated. The document is 211 x 274mm in size and is printed in red and black.

BUGATTI
LA 4L.900
TYPE 50 T

LE CHASSIS

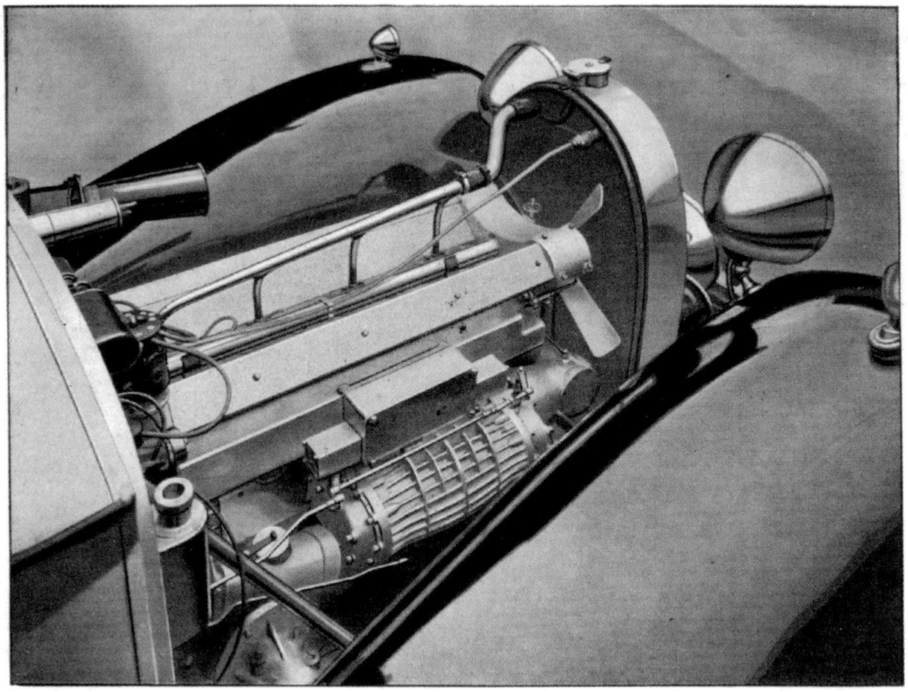

MOTEUR ET COMPRESSEUR

8 cylindres en ligne monobloc.
Alésage : 86 m/m ; Course : 107 m/m.
Deux arbres à cames en tête ; Compresseur toujours en prise.
Deux carburateurs automatiques.
Alimentation d'essence par pompe électrique.
Allumage par batterie, avec distributeur et bobine.
Les châssis sont livrés avec les appareils de contrôle suivants :
Compte-tours.
Compteur kilométrique.
Thermomètre d'eau.
Ampèremètre.
Manomètre d'huile.

L'ESSIEU avant creux (brevet Bugatti) est traversé par les ressorts. Ceux-ci se terminent à l'avant par des jumelles, et s'articulent à l'arrière sur un axe fixe. A droite le support arrière du ressort est constitué par un dispositif spécial, supprimant toute réaction dans les organes de direction sur les plus mauvaises routes. Les amortisseurs à friction sont commandés du tableau de bord pour l'avant comme pour l'arrière. Les roues d'aluminium coulé avec ailettes de refroidissement sont montées sur un moyeu cannelé et bloquées par un écrou unique. Le démontage de la roue met à portée de la main, les segments de frein, leur garniture et la clé de commande.

LES CARROSSERIES

CONDUITE INTÉRIEURE
1053

Cette carrosserie de forme très nouvelle, est le fruit d'expériences prolongées. Elle assure par son profilage outre une pénétration remarquable d'où découle vitesse plus élevée et moindre consommation pour une même puissance, un confort absolu et une visibilité optimum. Prévue pour 4 places, la caisse groupe ses passagers au milieu du châssis, ce qui assure à la voiture un centrage parfait, et l'équilibre des formes se conjugue avec l'équilibre des charges pour une maniabilité hors de pair. (Garniture cuir fin, verres de sécurité, pare-brise, vaste malle arrière complètent l'équipement de cette carrosserie.)

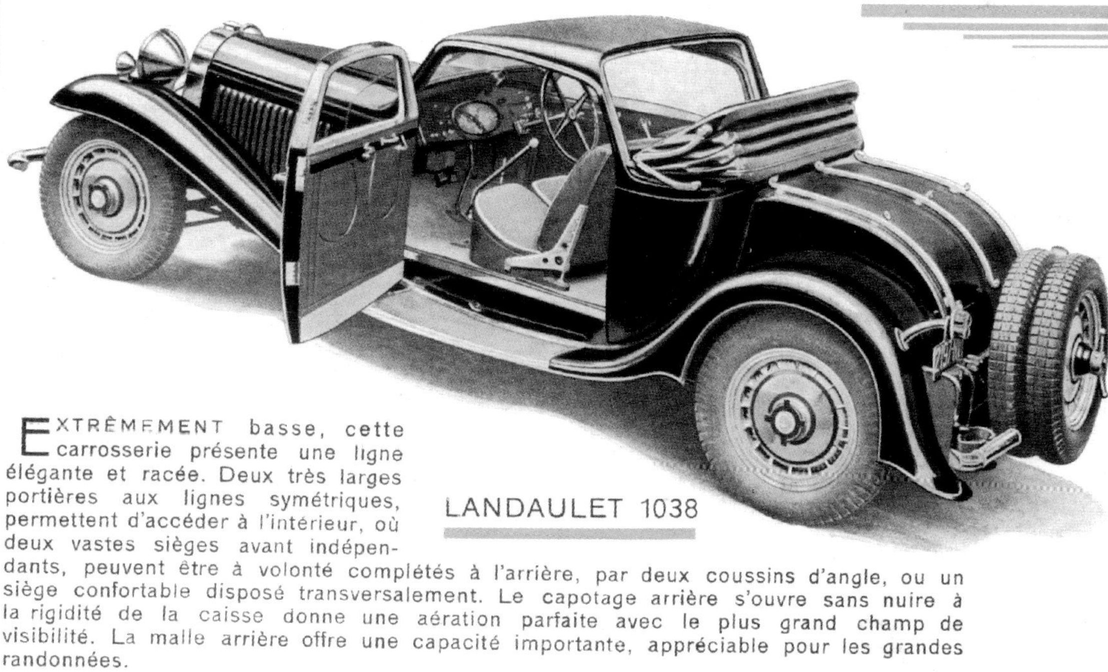

LANDAULET 1038

Extrêmement basse, cette carrosserie présente une ligne élégante et racée. Deux très larges portières aux lignes symétriques, permettent d'accéder à l'intérieur, où deux vastes sièges avant indépendants, peuvent être à volonté complétés à l'arrière, par deux coussins d'angle, ou un siège confortable disposé transversalement. Le capotage arrière s'ouvre sans nuire à la rigidité de la caisse donne une aération parfaite avec le plus grand champ de visibilité. La malle arrière offre une capacité importante, appréciable pour les grandes randonnées.

LANDAULET 1042

Vaste et confortable, d'accès facile grâce à 2 larges portières, cette carrosserie, comporte à l'avant, 2 sièges indépendants, et à l'arrière une banquette permettant de recevoir 3 passagers. Entièrement garnie de cuir de 1er choix, et le pavillon doublé de drap fin, cette caisse, grâce à la rotonde décapotable, donne principalement aux places arrière, une aération parfaite en même temps qu'une visibilité très étendue, Une malle arrière d'amples dimensions, des pare-choc avant et arrière, des feux de position sur les ailes avant, et deux avertisseurs couplés au bas du radiateur complètent cette carrosserie de grand style.

VUE DE DÉTAIL DE LA CAPOTE

Cette capote, limitée à la rotonde arrière présente l'avantage considérable de ne diminuer en rien la rigidité de la caisse, ce qui est une garantie de silence ; de plus chaque articulation est munie d'une bague de caoutchouc; enfin, l'étanchéité est obtenue grâce à la prévision sur tout le pourtour de la capote d'une bande de caoutchouc souple sur laquelle appuie un tube de caoutchouc épais.
Il faut noter l'important dégagement de la vue aux places arrière lorsque la capote est rabattue.

IMP. ARTS GRAPHIQUES – NANCY.

The title page, foreword and sample pages from the Bugatti London Depôt sales catalogue of 1932. The 12-page loose-leaf catalogue features a range of Bugatti models and is illustrated by colour plates stuck into the frames on individual pages. The brochure pages measure 287 x 240mm. Four more pages are reproduced overleaf.

ETTORE BUGATTI
AUTOMOBILES

BUGATTI

Head Office and Service Depôt
1-3, BRIXTON ROAD, LONDON, S.W.9

Telephone RELIANCE 3165 - 3166 *Telegrams*: BUGATTIMO, CLAP-ROAD, LONDON.

Foreword

1898 to 1932

THIRTY-FOUR years have now elapsed since the introduction of the first Bugatti car, since when untiring efforts, striving continuously year by year to improve the breed, have resulted in the production of the present day Models. Heading always towards the goal of perfection, we now confidently recommend the Bugatti car to the consideration of the discriminating motorist whatever be his needs.

In the minds of the "old time" motorist, the Bugatti is frequently looked upon as the motor-car built essentially for speed, and as such, requiring special attention and tuning to keep it up to concert pitch, also with chassis dimensions allowing for only scanty coachwork. In the past, up to a point, this view was justified, but TIMES HAVE CHANGED. Since the introduction of the present-day current Models, in the BUGATTI the public is offered a general purpose car which has silence, flexibility, smoothness, wonderful acceleration and highly efficient brakes, coupled with a top gear performance of from a crawling pace to a very high maximum speed. We therefore claim that the present-day BUGATTI car is equally suitable as a town carriage under congested traffic conditions, or for really high speed touring. Further, we particularly wish to emphasize that the modern BUGATTI requires no more servicing than any other standard motor car of repute.

The high efficiency as regards performance, always expected when the name BUGATTI is mentioned, has been fully maintained. An important feature which we wish to draw special attention to is, that the chassis dimensions to-day are such, that spacious, comfortable, high-class coachwork of the latest design is available as standard on all Touring-Town Chassis.

Proof is more convincing than mere words, for this reason we welcome being afforded the opportunity of demonstrating the Modern Bugatti to any interested motorist in whatever direction his requirements lie.

Ettore Bugatti

BUGATTI

8 CYLINDER 5 LITRE TYPE 46 CHASSIS FITTED WITH OCCASIONAL 4-SEATER "DROP-HEAD" COUPÉ.

NO engine can really be described as flexible that is not of the multiple cylinder type—and the 8-cylinders-in-line is now generally accepted as the ideal unit. This type was introduced at an early date in the Bugatti car and, from that early beginning, through years of experience and experiment, the perfect Bugatti engine of today has slowly evolved. The Designer has constantly aimed at two ideals; freedom from vibration and lightness of moving parts to allow high engine speed without apparent effort. The tremendous pulling power of the Bugatti Engine on top gear on a steep up-gradient is the measure of their success.

8 CYLINDER 5 LITRE TYPE 46 CHASSIS FITTED WITH 4/5-SEATER "SPORTSMAN" SALOON.

WHETHER his car be large or small, it is of vital importance to the driver for the steering to be safe, light in operation and free from excessive vibration. The Bugatti steering is designed to eliminate as far as possible all traces of road wheel shock. The steering wheel is of the flexible type, the steering column is supported in a rubber-bushed housing and a flexible leather coupling is inserted between steering column and steering box. This ensures driving comfort at high speeds and over bad surfaces, and the owner-driver of a Bugatti can count on the minimum of fatigue from the longest day's touring.

CURRENT PRICE LIST

8-cylinder 25.7 H.P. (3300 c.c.) TYPE 49 Chassis	£625.
Standard 3/4-seater "GRAND-SPORT" Coachwork on TYPE 49 Chassis	£150.
Standard 4-seater TOURER Coachwork on TYPE 49 Chassis	£225.
Standard 2-door "SPORTSMAN" COUPE or "SPORTSMAN" SALOON Coachwork on TYPE 49 Chassis.	£285.
Standard 4-door SALOON Coachwork on TYPE 49 Chassis	£340.
○ ○ ○	
8-cylinder 32.5 H.P. (5350 c.c.) TYPE 46 Chassis unsupercharged	£975.
8-cylinder 32.5 H.P. (5350 c.c.) TYPE 46 Chassis supercharged	£1175
Standard 4/5-seater TOURER Coachwork on TYPE 46 Chassis	£250.
Standard 4/5-seater 2 or 4-door "SPORTSMAN" SALOON Coachwork on TYPE 46 Chassis	£475.
Standard 5- or 7-seater 4-door SALOON Coachwork on TYPE 46 Chassis	£550.
○ ○ ○	
8-cylinder 30.6 H.P. (4900 c.c.) Supercharged TYPE 50 Chassis	£1725.

○ ○ ○

Estimates and Designs of Coachwork to special requirements supplied on demand.

○ ○ ○

Note:—Prices subject to alteration with fluctuation in the Rate of Exchange.

CONDITIONS OF SALE

PRICES The manufacturer's products will be supplied at prices current at the time of delivery. The Manufacturer reserves the right to revise his prices without notice, and orders are accepted only on that condition. The prices specified are subject to above conditions.

MODIFICATION The Manufacturer reserves the right to alter or modify without notice the specification or design of his cars as may be considered necessary.

DELIVERY Delivery of chassis or cars, spare parts and renewals, is in all cases at the Manufacturer's Depot, and every effort will be made to give delivery on the estimated date. The Manufacturer will accept no responsibility in respect of any delay in delivery or any claim arising therefrom.

PAYMENT All chassis or cars must be paid for prior to despatch from the Manufacturer's Depot, and, on leaving, such cars are at the purchaser's risk.

GUARANTEE All cars or chassis of the Manufacturer shall be accepted by the purchaser subject only to the following express warranty which shall exclude all conditions, warranties, representations and liabilities whatsoever (whether statutory or otherwise) which might exist but for this provision.

In the event of any defect being disclosed in any part of the car (except tyres and/or coachwork and/or accessories and/or electric fittings for which the Manufacturer shall not be responsible), the Manufacturer undertakes, on return of the defective part, carriage paid to his Depot, properly labelled with the chassis number and the owner's name and address, within twelve months after the delivery of the car, to examine it, and should any fault be found on such examination by the Manufacturer, due to defective material or workmanship, the Manufacturer will repair the defective part or despatch a new one in place thereof, carriage forward, free of charge.

GUARANTEE—*cont.* In the case of any fault being disclosed in the course of racing, or primarily due to wear and tear, dirt, misuse, or neglect, or in any chassis or cars which have been altered after leaving the Manufacturer's Depot, or which have been let out on hire or the identification numbers or marks on which have been removed or obliterated, no liability of any sort or kind shall be on the Manufacturer.

Chassis supplied by the Manufacturer are intended to be fitted with bodies of the same weight and general character to those shewn in the Manufacturer's catalogue, and if fitted with bodies differing therefrom, no liability of any sort or kind shall be upon the Manufacturer.

RISKS While customers' cars are in the Manufacturer's possession all precautions are taken against fire or damage from any cause, but he is unable to accept any liability in this respect, and will be glad if customers will note their cars can only be driven by members of the Manufacturer's staff at customers' risk and responsibility.

In order to avoid any misunderstanding, unless the manufacturer has written instructions to the contrary, the delivery of the car to him for testing or repair, with or without the owner's driver, will be taken as granting permission to drive the car on the above-mentioned conditions.

EXHIBITS All goods are sold upon the condition that they are not to be exhibited, without the previous consent in writing of the Manufacturer, at any exhibition, and that they are not re-sold when new at other than the full retail price as published in the Company's current price list.

SERVICE We have built up a Bugatti Service Organisation which enables us to offer adequate assistance to Bugatti owners who require anything done in the way of adjustments, overhauls or repairs Although the Bugatti revels in hard work, it repays careful and intelligent examination periodically.

All communications concerning adjustments, overhauls, repairs or the supply of replacements of damaged parts, should be addressed to

ETTORE BUGATTI
AUTOMOBILES

1—3 BRIXTON ROAD, LONDON, S.W. 9.

Telephone: RELIANCE 3165 (2 lines). *Telegrams:* "BUGATTIMO CLAPROAD LONDON."

Two pages from a twenty-page German-language brochure featuring the whole range of Bugatti models. The brochure, which measures 160 x 227mm, is printed in blue and black but is not dated.

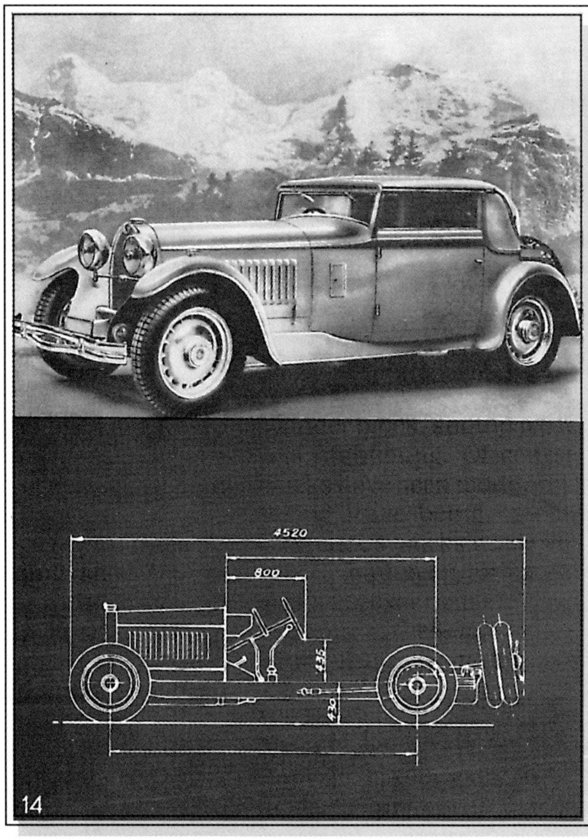

CHASSIS

4,900 LITER SPORT TYPE 50
MIT KOMPRESSOR

MOTOR: 8 Cylinder in Linie, Monoblock
BOHRUNG: 86 mm, HUB: 107 mm, CYLINDERINHALT: 4900 cm³
KURBELWELLEN-Lager direkt am Cylinderblock
2 NOCKENWELLEN, 2 VENTILE PRO CYLINDER
BATTERIEZÜNDUNG
KOMPRESSOR direkt durch den Motor angetrieben
TROCKENLAMELLEN-KUPPLUNG
GETRIEBE in der Hinterachse angeordnet
VORDERACHSE mit rundem Querschnitt aus Edelstahl geschmiedet
RECHTSSTEUERUNG
ALUMINIUMRÄDER (Patent Bugatti)
BEREIFUNG: 32×6,50
VIERRADBREMSEN mechanisch mit Ausgleichgetriebe
FEDERUNG: vorne ½ elliptisch, hinten ¼ elliptisch (Patent Bugatti)
ELEKTRISCHES LICHT UND ANLASSER (getrennte Apparate)
RADSTAND: 3 m 10
KAROSSABLE LÄNGE: 2 m 20
SPURWEITE: vorne und hinten 1 m 40
GEWICHT DES CHASSIS: ca. 1300 kg

15

A Bugatti advertisment featuring Chiron with T50 and emphasising the marque's virtues.

APPENDIX I
THE FACTORY SALES RECORDS

The Type 46 – reported to be EB's favourite model – can be considered a successful model, with about four hundred & sixty seven cars built. The picture of EB, family and staff standing around the 500th car built is patently false, but it could be chassis number 46500.

Results were good, especially taking into account the fact that Bugatti was considered a builder of noisy and oily sports racing cars and this model was aimed at the luxury town carriage market.

Sales, although principally to the French market, were also good in Switzerland, and Belgium: Sweden, Germany, Italy and Holland produced buyers as well. The London Depôt managed to dispose of around twenty four cars, which can be considered good, taking into account the economic climate at the time.

Eight chassis were invoiced to London on 31 December 1930, possibly for fiscal reasons.

Towards the end of production conditions became very difficult and one or two chassis were sold to agents at knock-down prices. Even a racing driver - Louis Charavel - who raced under the pseudonym 'Sabipa', took two which, in 1968, were discovered still in brand new condition and without bodies.

The car was certainly expensive with a British list price of £975 for the chassis, whilst a Mulliner sports saloon carried a list price of £1525. The D8 Delage, with a Chapron sports saloon body, was listed at £1090 and the chassis price of the well-established 4.5-litre Bentley was £1050.

Unfortunately, surviving records for the Type 46 are grossly incomplete and it is hoped that this book will encourage owners and enthusiasts to fill in some of the missing links.

Type 50 production totalled sixty five cars and it was a remarkable feat that buyers were found, considering it was an ultra-expensive and complicated car which fell between the categories of large sports and luxury cars.

The first car was delivered in October 1930, after which twenty six more were sold in 1931, ten in 1932 and twenty five in 1933. Two final sales were recorded in 1934.

No attempt was made to sell the Type 50 in Great Britain and it was not even listed in Motor Show guides of the period. However, well-known Bugatti enthusiast Sir Robert Bird purchased chassis number 50118, which must have cost in excess of £2000, and Lord Cholmondeley bought one direct from the Paris showroom; it is believed he kept the car in France.

The entire production of Type 50s was sold to French and Swiss customers, the latter country taking eight cars via Bucar of Zurich, one of the most succesful Bugatti agents.

The Type 46 and 50 probably suffered some sales resistance from agents due to servicing difficulties; it is a fact that work on the valves necessitated a complete engine stripdown. But then, Bugatti would argue, quite reasonably, that, by the time the valves required attention, the rest of the unit would require dismantling anyway for a thorough inspection and overhaul as necessary. The fact is the cars were built on sound, if unconventional, lines and were very reliable and possessed of a long life.

Type 46. A well-known photograph of Ettore Bugatti, family and works personnel outside the Bugatti château. The car is widely quoted as being the 500th Type 46 but, as only about 467 were ever constructed, it's probably chassis number 46500.

CH.	ENG.	AGENT	INV.	OWNER	STYLE	COACHBUILDER	REMARKS
46121		Christy, Lyon	16.11.29		Cabriolet		
46122				Brulé			
46123		Friderich, Nice	6.12.29		Coach		
46124							
46125				Shakespeare Schlumpf	DH Coupé	Letourneur & Marchand	Wire wheels
46126					Saloon		
46127		Remiarty, Liége	24.1.30		Conduite Int.		
46128		Muller & Cluzel, Mulhouse	7.11.31		Coach		
46129							
46130		Londow & Wilson Paris	30.11.29		Berline		Delivery to Paris by Williams.
46131	11	Londow & Wilson Paris	1.2.30		Chassis		
46132		Marcel Laroob, Héricours	29.11.29		Cabriolet	Letourneur & Marchand	
46133	8					Bugatti	
46134			30.11.29	Chiron	Tourer	Brooks of Biarritz	At works 25.10.30, delivery to Brooks, Biarritz.
46135		Rigal, Paris	17.12.29	Recquart/Grell/Von Raffay	7-seater saloon	Gangloff	Returned works for repairs, left 12.29. Works 2.31. Delivered Vizier.
46136							
46137	17						
46138		Larousse			Chassis		
46139		Petrovits, Budapest			Berline		
46140							
46141			Dec 29				
46142	22						
46143							
46144							
46145							
46146							
46147							
46148							
46149							
46150		Sweden	17.1.30		Chassis		
46151		London	16.1.30				
46152		Paubrei, Pars	19.12.29		Chassis		
46153	35	Casteé, Pars	2.2.31			Letourneur & Marchand	Del. with Touriste body
46154							
46155			16.4.30	Maurice Rost	Tourer		Works demo. Delivery to Tourcoing.
46156							
46157	27	Lehmann, Paris	13.8.30		Touriste		
46158	74 or 14	De Rowan Paris	6.2.30		Chassis		
46159							
46160							
46161							
46163	195	London Depot	2.4.30	Montmorensi	2-dr Sal	Weymann	Body semi-panelled. Eng. chgd. 2nd owner J. McCarric.
46164		Monestier					
46165							
46166	47						
46167		Bordeaux	27.1.30				Chassis to Paris coachbuilder
46168							
46169		Divo, Paris	3.7.30		Touriste		

CH.	ENG.	AGENT	INV.	OWNER	STYLE	COACHBUILDER	REMARKS
46170	42	Detoren (chassis only)		Lanfanay			
46171				Chateau de Frandson			
46172		Bucherer, Bonn	16.7.30			Letourneur & Marchand	
46173		Remiarty, Liége			Chassis		
46174		Renoyer				Letourneur & Marchand	
46175		Dr Kocher, Falience	20.1.30		Chassis		
46176							
46177							
46178	52					Letourneur & Marchand	
46179		Wilson Mondow?	16.1.30		Chassis		Delivery by Bouriat
46180							
46181	65						
46182		Monestier					
46184	93	London	25.2.30	Turner			
46185		Rovisser, Rémirecours			Chassis		
46186	59						
46188	78		1930	Schlumpf	6 light Saloon	Gangloff	Wire wheels
46189							
46190							
46191		Foltlen, Paris	2.4.30		Chassis		
46192	81?						
46193							
46194	122			Valetta			
46195		Fayol, Vichy	8.2.30		Chassis		
46196							
46197		Motte, Tourcoing	25.3.31		Saloon		
46198		Monestier					
446199	139			Labbé			
46200		Mondow, Paris	7.2.30		Chassis		
46201							
46202	60						
46203							
46205							
46206		Monestier					
46207		Mondow, Paris	8.2.30		Chassis		
46208	77			Bardin Raben-Lavetzav,			
46209	44?	London, via Larousse, Paris	16.1.30	Denmark			
46210	85	Rigal, Paris	11.2.30		Chassis		
46211	31	Mondow, Paris	11.2.30		Chassis		
46212							
46213		Monestier	June 1930				
46214	73?						
46215	83						
46216		H.Esdier, Bâle	21.4.30		Touriste		
46217							
46218	113	Kathou, Tourcoing	10.3.30	Hucke/Southward			Badly damaged by fire in New Zealand 1992.
46219	54	London Depot	16.1.30	Eric Giles	2-dr Sal	Weymann	Later owners: Scott-Moncrieff/Giles/Wilkinson/Cooke-Leicester
46220							
46221	91						

46222		Omnia, Munich	11.4.30		Chassis		
46223							
46224	90	Monestier					
46225		Monestier					
46226				London Depôt		Weymann	
46227		Regal, Paris	3.4.30		Chassis		
46228							
46229					FH Coupé	De Villars	
46230		Omnia, Munich	8.8.30		Chassis		
46231		Mathon, Tourcoing			Chassis		
46232	110	London	6.3.30	Harrison			
46233							
46234		Blanc & Gauche	23.4.30		Chassis		
46235	94						
46236		Mme Coty, Paris	17.4.30		Berline		
46237		Bucar, Zurich	28.4.30				
46238							
46239							
46240							
46241	185	London Depôt	10.5.30		2-dr Sal. (Coach)	Gangloff	Std Bugatti design. Cheshire, England, license no. LG 4679
46242	109						
46243		Basurel, Roubaix	21.12.31		Saloon		
46244	155						
46245	71	Mondow, Paris	1.3.30	Von Raffay	Touriste?		
46246							
46247							
46248							
46249	114						
46250				Camille Bourges			
46251							
46252							
46253							
46254							
46255							
46256	68?	London	17.4.30	Baron D'Erlanger	Berline		
46257		London	17.5.30				Brooks auction
46258		Pescuti-Laraca, Milan	30.7.31		Torpedo		
46259							
46260							
46261							
46262		Aurdux, Roanne	17.4.30		Chassis		To Paris coachbuilder
46263							
46264							
46265		Onnicium? Nancy	5.3.30		Chassis		
46266							
46267							
46268							
46269				McBean			
46270							

CH.	ENG.	AGENT	INV.	OWNER	STYLE	COACHBUILDER	REMARKS
46271		Geoffroy, Troyes	7.3.30		Chassis		
46272							
46273		Lamberjack, Paris	26.3.30		Chassis		
46274							
46275	128						
46276							
46277							
46278				Ruch Weidmoller		Gangloff	Engine in 46533.
46279							
46280		Moudin & Giradot, Paris	27.3.30		Chassis		
46281							
46282							
46283							
46284		Monestier, Lyon	29.4.30		Chassis		
46285							
46286							
46287				Shakespeare Schlumpf.			Postwar body built in USA, alloy wheels.
46288		Rouriat, Paris		Bob Baer			
46289							
46290							
46291							
46292							
46293				Heusboek			
46294							
46295							
46296		Laroussé, Paris	5.7.30		Touriste		
46297							
46298							
46299							
46300							
46301							
46302							
46303							
46304							
46305		Lamberjack, Paris	2.5.30		Chassis		
46306							
46307		Laroussé, Paris	16.5.30		Chassis		
46308							
46309							
46311		Figncron, La Rochelle	5.5.30		Chassis		
46312							
46313		Laroussé, Paris	12.7.50		Chassis		
46314							
46315							
46316		Bucar, Zurich	3.5.30		Chassis		
46317							
46318		Figraut Escla?	28.5.30		Chassis		
46319		Bucar, Zurich	3.5.30		Chassis		
46320							

46321							
46322		Monestier					
46323		Nauvel-Bize, Paris	13.12.30		Cabriolet		
46324		Chiron, Paris	29.9.30		Roadster		
46325		Omnium, Nancy	12.9.30	Schlumpf			Engine in 50157.
46326							
46327		Bucar, Zurich	11.6.30			Gangloff	Chassis to Colmar so Gangloff assumed coachbuilder.
46328		Fiddler, Paris	4.7.30		Cabriolet		
46329		Laroussé, Paris	26.5.30		Chassis		
46330							
46331	199	Mondau, Paris	17.5.30	Coombes	DH Coupé	Figoni	Auctioned by Poulain-le-Fur, 1993 - now Lord Raglan
46332							
46333							
46334		Lamberjack, Paris	20.5.30		Chassis		
46335							
46336		Monestier					
46337							
46338		Bucar, Zurich	19.6.30			Gangloff	Chassis to Colmar so Gangloff assumed coachbuilder.
46339							
46340	368			Ten Cate, Holland			Second body 'basket work' 2-seater.
46341		Gaddet, Paris	4.6.30		Roadster		
46342							
46343		Mondau, Paris	7.6.30		Chassis		
46344							
46345	98						Engine now in 46491.
46346							
46347		Lamberjack, Paris	4.6.30		Chassis		
46348		Hermorel, Villefranche	4.6.30		Chassis		
46349							
46350		Lamberjack, Paris	20.6.30		Chassis		
46351							
46352							
46353							
46354	233	London	1.7.30	Broomfield			Second body 2-seater GT type.
46355	278			Prince Hohenzowe			Engine ex-46424.
46356		London	28.6.30				
46357		Mondau, Paris	21.6.30		Chassis		
46358							
46359		Lamberjack, Paris	20.6.30		Chassis		
46360				C.Canzous, Greece			
46361		Luing, Paris	30.6.30		Chassis		
46362		Peigues, Molsheim	24.10.30				Chassis delivered to Mulhouse (Gangloff?)
46363	240			Prince Ruspoli	Saloon		
46364							
46365				Mette			
46366							
46367							
46368							
46369							

CH.	ENG.	AGENT	INV.	OWNER	STYLE	COACHBUILDER	REMARKS
46370							
46371							
46372							
46373							
46374				Marguet			Auctioned Dec. '81.
46375	249	Mondau, Paris	12.7.30	Graf	2-dr Sal. (Coach)	Weymann	
46376							
46377		Figreud, Escle	31.10.30		Chassis		
46378							
46379		Sybbelied, Rheims			Chassis		
46380							
46381		Van Ramshoäs		Schattelear, Germany			
46382							
46383							
46384							
46385							
46386		London	24.7.30			Weymann?	
46387		London	24.7.30			Weymann?	
46388		Mondau, Paris	19.7.30		Chassis		
46389							
46390							
46391							
46392							
46393							
46394							
46395							
46396							
46397							
46398							
46399							
46400		Jans, Lyon	22.7.30		Chassis		
46401							
46402							
46403							
46404	259	London	25?.7.30	Black			
46405	269	Paris	Feb 36	Tuttle	FH Coupé	Gaston Grummer	
46406							
46407	272			Cornu	FH Coupé		
46408							
46409							
46410		Lagnice, Algeria	11.10.30		Cabriolet		
46411							
46412							
46413							
46414							
46415							
46416							
46417							

46418	312			Heubult			
46419		Omnium, Écrival	20.9.30		Chassis		
46420							
46421		London	17.9.30				
46422		Schlumbereer Fuebwiller			Chassis		
46423							
46424	278						Engine in 46355.
46425							
46426		London	17.9.30				
46427							
46428							
46429		Causson-Carron, Brussels	8.10.30		Chassis		
46430							
46431							
46432							
46433	292			Hicguel		Van Vooren	
46434	294		Built 1933	Schlumpf	Chassis		Bought direct from Bugatti (prewar).
46435	303	Friderich, Nica	14.10.30	Augustin	DH Coupé	Van Vooren	
46436		Friderich, Nice	14.10.30		Chassis		
46437			Jan 1931				
46438			Jan 1931				
46439							
46440							
46441							
46442							
46443							
46444		Monestier	Mar 1932			Van Vooren	
46445							
46446							
46447							
46448							
46449	368	H. Nilleir?	4.4.36			Van Vooren	
46450							
46451							
46452		Van Ramshorst	1931			Rhodins, Amsterdam	
46453							
46454							
46455		Van Ramshorst	1936		DH Coupé		
46456							
46457							
46458		Coussin, Erussels	24.11.30		Chassis		
46459							
46460							
46461							
46462							
46463							
46464							
46465							

CH.	ENG.	AGENT	INV.	OWNER	STYLE	COACHBUILDER	REMARKS
46466							
46467							
46468							
46469		Charles Faroux, Paris	6.1.31		Saloon		
46470	372	Rost, Paris	18.3.31	T. Waugh	Saloon		
46471		Monestier	Mar 1931				
46472							
46473							
46474							
46475							
46476							
46477							
46478							
46479							
46480							
46481							
46482	327			Sold by EB's executors	Coach Profilée (2nd body)		Later owners: Richardson/Rhealt/Schlumpf. Alloy wheels.
46483							
46484	334						
46485	298						
46486							
46487	336						
46488	309						
46489	316			Schlumpf			Bought new (postwar).
46490	46S						
46491	345			Glass Charavel	Rep. Coach Profilée		New chassis stored France till 60s. New body UK '93.
46492							
46493							
46494							
46495		Lagnier, Algeria	15.7.31		Touriste		Shipped from Marseille.
46496		Labourdette, Paris	31.3.31			Labourdette	
46497							
46498							
46499							
46500		Omnium, Autos, Nancy	31.3.31		Chassis		
46501		Raffaeli			Roadster	Bugatti	
46502	398		24.4.36		Saloon	Chapron	
46503	417			Jack Delileux		Million Guiet	
46504							
46505		Monestier					
46506		Pescuti, Milan	17.10.31		Saloon		
46507							
46508							
46509							
46510	326						
46511	339						
46512				Glass?	DH Coupé	Gangloff	Meyer, Holland. Alloy wheels '88 for Peking-Paris race.
46513							
46514							
46515							

Chassis	Engine	Agent/Owner	Date	Later owner	Body	Coachbuilder	Notes
46517		Kelber	2933	Colombo			
46518							
46519							
46520							
46521							
46522				Kreyer 1979			
46523		Bucar, Paris	26.10.31*	Shakespeare Schlumpf	4 Light Saloon	Million Guiet	Car finished & sold 1933. * Chassis delivery date.
46524	448	Hans Schmidt, Stockholm	20.10.31	Precision Cars		Touriste	
46525	399			Roth			
46526							
46527							
46528		Bucar, Zurich	28.8.31				Chassis delivered to Colmar (Gangloff?)
46529	358?	London	30.12.30				
46530	378?	London	30.12.30				
46532	381?	London	30.12.30				Later owners: Margulies (1967)/Hucke
46533	382	London	31.12.30	Williamson-Hancock	4 light Saloon	Freestone & Webb	Engine now in 46278?
46534	383	London	31.12.30	Clutton	4 light Saloon	James Young?	
46535	185	London	31.12.30	Virr			
46536	387?	London	31.12.30				
46537				Hucke?			
46538		Van Ramsho st	Nigos				German body.
46539							
46540	46S						
46541							
46542							
46543		Bucar, Zurich	22.7.31				Chassis delivered to Colmar (Gangloff?)
46544							
46545		Bucar, Zurich	9.9.31				Chassis del. to Colmar (Gangloff?)
46546	186			Harrah			
46547							
46548							
46549							
46550		Castelbarco, Milan	17.10.31		Cabriolet		
46551							
46552	46S			Schlumpf			
46553	390	London	23.9.31	Tremlet			
46554	391	London	23.9.31				
46555	414?		1934	Schlumpf		Dolce of Cannes	Wire wheels.
46556	400?	Casa Mayou, Paris	19.10.31				Chassis delivered to Colmar (Gangloff?)
46557							
46558							
46559							
46560	1S			Schlumpf	4 Light Saloon	Gangloff	Alloy wheels. Stored in Marseilles as new car till '60s.
46561							
46562	S	Monestier, Lyon	1.12.31		Saloon	Million-Guiet	
46563	S						
46564	S						
46565							

CH.	ENG.	AGENT	INV.	OWNER	STYLE	COACHBUILDER	REMARKS
46566							
46567							
46568							
46569							Engine only owned by Bhibau.
46570	49	London		Albright			Now 'S'
46571		London	1.4.32				Engine now in 46232
46572	S						
46573	S						
46574	197S			Schlumpf	4 Light Saloon	Gangloff	
46575	S						
46576							
46577		London	13.4.32	Pengilley Col Giles			
46578							
46579							
46580	351/123	Paris	Nov. 1933	Ruini			
46581	S	Monestier	1933				
46582	S						
46583	10S	Zurich	21.5.32	Kreyer			
46584	S						
46585	9S	Monestier	Apr. 1932			2-door saloon	Billeter & Cartier, Lyon Later owner Carter of Stevenage.
46586	S						
46587	16S	London Depot	23.6.33	Hampton Burnett	FH Coupé	James Young	Later owners: Fawcett/Hampton/Burnett.
46588	12S	London Depot	23.6.33	Norton	DH Coupé	French body?	Road registered London 1934 (BXO 725).
50112		Guy Bourriat, Paris	21.10.31				
50113		Lamberjack, Paris	21.10.31				
50114		HRH Prince del Drago, Paris	10.11.31				
50115		Fayol, Vichy	13.3.31				
50116	1	Lamberjack, Paris	11.10.30				Later owner: Stanley, USA. New Roadster body Peter Agg 1991.
50117		Lamberjack, Paris	2.2.31				
50118	7	Sir Robert Bird, London	3.2.31				
50119		Lamberjack, Paris	10.2.31				
50120		Achille Fould	1931				
50121		Bucar, Zurich	1.12.30				
50122		Bucar, Zurich	10.12.31				
50123		Derain, Paris	30.3.31				
50124	35		1932?				
50125		Bucar, Zurich	11.4.31				
50126	12	Bucar, Zurich	10.4.31				
50127		Marion, Lyon	26.5.31				
50128		Bucar, Zurich	10.4.31				
50129		F.Baudry, Cernay	12.5.31				
50130		Garage Besion, Béziers	30.10.31				Chassis survives.

50131		Guy Bourriat, Paris	30.10.31			
50132		Pontarlier ?	1.8.33			
50133		Léon Fabrè, Paris	17.2.32			
50134		Lamberjack, Paris	16.6.31			
50135		Bucar, Zurich	7.7.31			
50136		St Didier Auto, Paris	13.7.32			
50137		Bucar, Zurich	7.7.31			
50138		Bucar, Zurich	7.7.31			
50139		Lamberjack, Paris	15.7.31			
50140		Lamberjack, Paris	28.7.31			
50141		Lamberjack, Paris	2.9.31			
50142		G.Dubois, Paris	16.9.31			
50143	147	Antoine Schumann, Paris	4.12.31			
50144		Bartifaut, Paris	17.2.32			
50145		Lamberjack, Paris	8.11.32			
50146		Astan Finaly, Paris	23.3.32			
50147		Louis Chiron, Paris	28.10.32			
50148		Ct. Czaykowski, Aubufillo	11.7.32			
50149		Auto, Nancy	26.8.32			
50150		Teillac, Paris	21.1.32			
50151		H.Morgan, Paris	29.11.33			
50152			30.3.33		50T	
50153		Evidal, Marseilles	4.12.33		50T	
50154		St Dizier	13.3.33		50T	
50155			25.9.33			
50156		Paris	22.1.33		50T	
50157		Duc de Valauzy, Paris	30.11.32		50T	Now a chassis with 46 engine.
50158	6	Chapelier, Paris	28.2.33		50T	
50159		Marion, Lyon	24.1.33		50T	
50160	2OT	Clermont Ferrand	24.1.33		50T	
50161	9	De Cystria, Paris	1.2.33		50T	
50162		Lamberjack, Paris	7.2.33		50T	
50163		Henzé Bounet, Paris	10.2.33		50T	
50164		Créquy et Carnette, Lille	22.3.33		50T	
50165		Créquy et Carnette, Lille	22.3.33		50T	
50166	17	Neuilly	13.10.33		50T	
50167	20	Paris	113.6.33		50T	
50168		Créquy et Carnette, Lille	13.6.33		50T	
50169		Garage Bayard, Paris	25.7.33		50T	
50170	23	Paris			50T	
50171	25T	Stapelle Gennevilliers	23.6.33		50T	
50172		Reims	31.7.33	Armandias, President Messier	50T	
50173		R. Bader, Paris	11.8.33		50T	
50174	22T	Monestier Lyon	26.6.33		50T	
50175		Le Surque, Paris	29.7.33		50T	
50176	43T	Grenoville Asniéres	5.12.34		50T	
50177	21					
50180		Works ca , then Schlumph			50B	Retained by works untill 1965.

BUGATTI 46/50

APPENDIX II THE BUGATTI TRUST

WHY A BUGATTI TRUST?

The motor cars built by Ettore Bugatti stand apart from all others in the manner in which they combine aesthetic appeal, efficient working and often highly innovative design.

They do, in fact, represent the zenith of the combination of art and engineering.

His approach and accomplishments, in many other fields as well as motor cars, make him an important 20th Century figure.

THE TRUST & ITS OBJECTIVES

The Bugatti Trust was created by a small group who have been Bugatti enthusiasts for many years. It is a registered charity whose objects are the establishment of a museum and research centre devoted to the historical and technical achievements of Ettore Bugatti, to promote knowledge of his works and, through this, an awareness of the importance of aesthetics in design.

GENIUS AT THE DRAWING BOARD: ETTORE BUGATTI 1881-1947

Ettore Bugatti expressed his genius with a pencil at his drawing board, outlining ideas which his small team of draughtsmen would turn into detailed production drawings.

He was a consultant designer before he was 21, being responsible for several different types of motor car. In 1909 he opened his own factory, producing successful designs under the Bugatti name.

These included a line of remarkable racing cars, of which the Type 35 was perhaps the outstanding example, and many high grade touring and sports models.

As well as cars he designed many of the tools with which they were built and did not restrict his work to the automotive field. Railcars, boats, aero engines, even a revolutionary aeroplane with twin coupled Bugatti engines and an all-wood fuselage came off his drawing board.

In everything Bugatti produced, at first on his own, and later with the help of his son Jean, he combined high grade engineering with attractive appearance.

This combination, essential to successful product design, gives the work of Ettore Bugatti a continuing relevance for contemporary engineering and industrial design, where performance excellence and aesthetic appeal are increasingly expected.

THE TRUST'S STUDY CENTRE

The Study Centre contains one of the world's most important collections of material on the work of Bugatti.

This includes original Bugatti designs and drawings, books, catalogues and thousands of photographs, many of them unique. And, although the Centre is not a car museum in the normal sense, there is space for the display of components, engines and a representative car or two.

Nobody could spend time in the Centre without becoming aware of how important aesthetics were to Bugatti, as well as good engineering design. This balance is apparent in all the works on show, and to increase appreciation of its importance is the primary educational purpose of the Centre.

Open to the general public, the Centre will no doubt attract many who have heard of Bugatti cars still rallying or racing, or being sold for enormous sums. They will find much to interest them, and perhaps among them there will be some young people who are stimulated to take up a design career in art or engineering.

The Centre's other role is to provide for the serious study of the materials on display or in the library by students of design or technology, authors and journalists. Facilities exist to make available prints of the photographs at the Centre - certainly the largest collection of Bugatti photographs in existence.

The Centre is housed in a special building located at Prescott, near Cheltenham, England, on a site made available by the Bugatti Owners Club, the oldest of the several Bugatti Clubs throughout the world.

ON DISPLAY

Always at the heart of the display there will be a Bugatti car, the type and model of which will vary from time to time.

In addition there is an absorbing collection of other Bugatti exhibits, with a wide appeal both to the expert and the general public alike.

Presented on large display panels is historical data on Ettore Bugatti and his family, including his father Carlo, a furniture designer, his sculptor brother Rembrandt and son Jean.

- Illustrations of all Bugatti's car designs from 1900 to his death
- A display of the complete collection of Bugatti catalogues. Some of these are of great aesthetic, as well as monetary, value
- Sectioned assemblies including an example of Bugatti's first (1922) production 8-cylinder engine
- Examples of engine and gearbox components, cast wheels and radiator shapes
- Data on Bugatti's coachwork, railcars, aero engines, aeroplanes, boats and yachts, horsedrawn carriages
- Bugatti the Inventor - with 1,000 patents
- Bugatti the Engineer - the good, and the less good
- Bugatti the Man - his horses and other personal details
- Bugatti's son Jean - his contribution, especially in coachwork
- A library of books on Bugatti and related subjects
- A photographic library, with many photographs displayed and over 2,000 accessible for study
- There is also a shop offering Bugatti books, document reproductions and other memorabilia

A CALL FOR MEMBERS

The Trust needs more members if it is to expand its work, and bring home to more and more people, especially the young, the lesson that Bugatti teaches - good engineering must be married to good aesthetic design, for a product to be successful in the highest terms.

This lesson, increasingly important to designers and engineers as expectations rise worldwide, is central to the educational work of the Trust. You could be confident of contributing to this philosophy by becoming a member, with the advantage to yourself of being able to use the Study Centre without charge. You would also receive an informative quarterly newsletter.

If you would like to become a member of the Trust, please write for an application form to:

The Bugatti Trust, Prescott, Gotherington, Nr Cheltenham, GL52 4RD, England.

BUGATTI 46/50

INDEX

Also from Veloce Publishing –

A comprehensive, radical look at the history and development of the Type 57 Grand Prix Bugattis. New material challenges traditional beliefs about these historic cars, and rejects some long-standing conventions. Myths are explored and truths are revealed in this book celebrating all aspects of these remarkable cars and their creators.

ISBN: 978-1-845847-89-0
Hardback • 24.8x24.8cm • £50* UK/$85* USA • 176 pages • 158 colour and b&w pictures

Reprint of a Veloce classic!
A concise history of the Bugatti Type 57 including the supercharged 57S and the Type 64 & 101. Packed with period illustrations, this is the standard reference on the 57 and its close relatives.

ISBN: 978-1-845848-71-2
Hardback • 25x20.7cm • £45* UK/$75* USA • 240 pages • 300+ colour pictures

For more info on Veloce titles, visit our website at www.veloce.co.uk • email: info@veloce.co.uk • Tel: +44(0)1305 260068
* prices subject to change, p&p extra

Dear Reader,
We hope you enjoyed
this Veloce Publishing
production.
If you have ideas for other
books on Bugatti - or any
marque - please write and
tell us.
Meantime, Happy
Motoring!

THE END